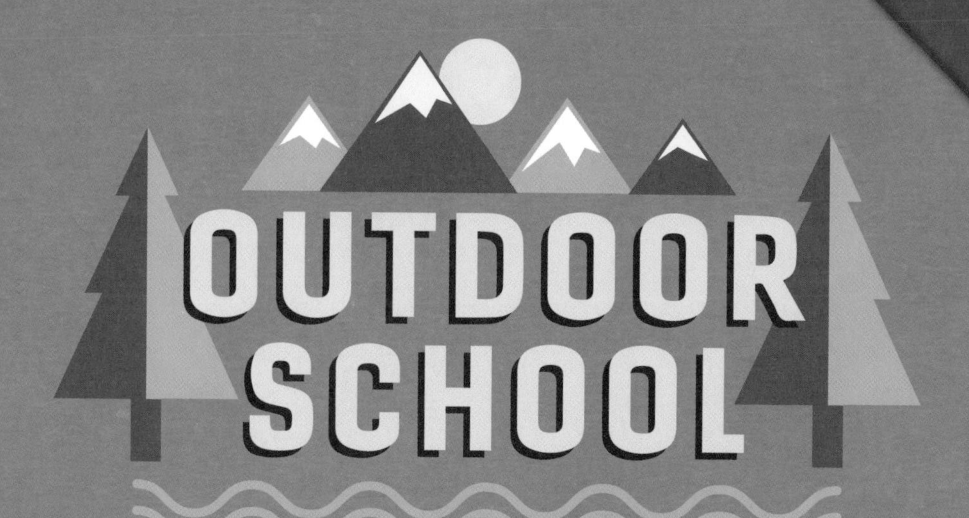

ANIMAL WATCHING

MARY KAY CARSON

ILLUSTRATED BY
EMILY DAHL

Odd Dot New York

An imprint of Macmillan Publishing Group, LLC
120 Broadway, New York, NY 10271
OddDot.com

Library of Congress Cataloging-in-Publication Data is available.
ISBN 978-1-250-23083-6

Cover Designers TaeWon Yu and Tim Hall
Interior Designers Carolyn Bahar and Christina Quintero
Colorist Guilia Borsi
Editor Nathalie Le Du
Animal Consultant Heather L. Montgomery

Illustration credits: Art from *Golden Guides: Mammals*, *Golden Guides: Birds*, *Golden Guides: Reptiles and Amphibians*, and *Golden Guides: Fishes by* James Gordon Irving. Art from *Golden Guides: Bird Life* by John D. Dawson. Red-shouldered hawk image from the Biodiversity Heritage Library. Contributed by Smithsonian Libraries. Snowy egret image from the Biodiversity Heritage Library. Contributed by University of California Libraries. Red bat image from the Biodiversity Heritage Library. Contributed by American Museum of Natural History Library. www.biodiversitylibrary.org

Printed in China by 1010 Printing International Limited, North Point, Hong Kong

First edition, 2021

1 3 5 7 9 10 8 6 4 2

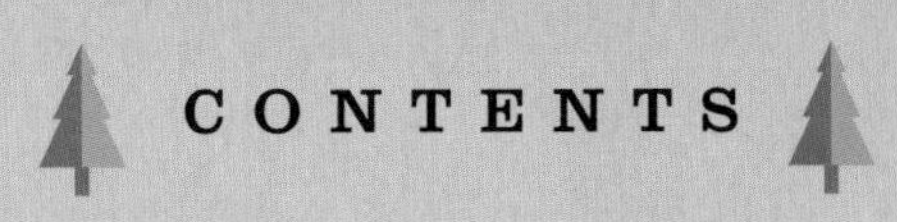

PART I: WHAT IS ANIMAL WATCHING? 1

Chapter 1: Meet the Locals 3
Chapter 2: Scout for Wildlife 7
Chapter 3: Become an Active Observer 16

PART II: BIRDS 24

Chapter 1: How to Be a Birder 27
Chapter 2: Natural Born Flyers 30
Chapter 3: Spot Birds by Silhouette and Shape 36
Chapter 4: Scout Birds by Size and Place 42
Chapter 5: ID Birds by Color 47
Chapter 6: What Birds Do 56
Chapter 7: How Birds Live 78
Bird Identification 86

PART III: MAMMALS 188

Chapter 1: How to Spot Mammals 191
Chapter 2: How to ID Furry Friends 202
Chapter 3: Mammal Tracking 215
Chapter 4: How Mammals Live 222
Chapter 5: What Mammals Do 232
Mammal Identification 240

PART IV: AMPHIBIANS AND REPTILES 302

Chapter 1: How to Find Herps 305
Chapter 2: How to ID Frogs, Toads, and Salamanders 315
Chapter 3: How to ID Snakes, Lizards, and Turtles 322
Chapter 4: Herp Tracking 330
Amphibian Identification 338
Reptile Identification 358

PART V: FISH 382

Chapter 1: How to Find Fish 385
Chapter 2: How to ID Finned Friends 393
Fish Identification 406
101 ACHIEVEMENTS 438
INDEX 440

OPEN YOUR DOOR.
STEP OUTSIDE.
YOU'VE JUST WALKED INTO
OUTDOOR SCHOOL.

Whether you're entering an urban wilderness or a remote forest, at Outdoor School we have only four guidelines.

- → **BE AN EXPLORER, A RESEARCHER, AND—MOST OF ALL—A LEADER.**
- → **TAKE CHANCES AND SOLVE PROBLEMS AFTER CONSIDERING ANY RISKS.**
- → **FORGE A RESPECTFUL RELATIONSHIP WITH NATURE AND YOURSELF.**
- → **BE FREE, BE WILD, AND BE BRAVE.**

We believe that people learn best through doing. So we not only give you information about the wild, but we also include three kinds of activities:

TRY IT → Read about the topic and experience it right away.

TRACK IT ↘ Observe and interact with the wildlife, and reflect on your experiences right in this book.

TAKE IT TO THE NEXT LEVEL ↗ Progress to advanced techniques and master a skill.

Completed any of these activities? Awesome! Check off your accomplishment and write in the date.

This book is the guide to the adventures you've been waiting for. We hope you'll do something outside your comfort zone—but we're not telling you to go out of your way to find danger. If something seems unsafe, don't do it.

Don't forget: This book is **YOURS**, so use it. Write in it, draw in it, make notes about your favorite waterfall hike in it, dry a flower in it, whatever! The purpose of Outdoor School is to help you learn about your world, help you learn about yourself, and—best of all—help you have an epic adventure.

So now that you have everything you need—keep going. Take another step. And another. And never stop.

Yours in adventure,

THE **OUTDOOR SCHOOL** TEAM

PART I

WHAT IS ANIMAL WATCHING?

What would YOU do?

You're hiking along an awe-inspiring mountain trail out West. What could be better! You've already spotted two species of bird you've never seen before, crossed a clear stream full of trout, and heard at least four kinds of squirrels calling. Wow! As you follow the trail out into a sunny meadow you suddenly see a big dark bear. It's on the trail up ahead! And the bear is looking right at you. What do you do? You know not to run or climb a tree—bears are better at both than humans. But is it a black bear? Should you yell, clap, and act tough? Or is it a grizzly bear? Would you know which bear it was? *What would you do?*

CHAPTER 1

Meet the Locals

Humans aren't alone on earth. We share the planet with millions of kinds of living creatures. No matter where you live, animals are around. Squirrels and spiders scamper across the grass you walk on, fish and frogs frolic in nearby streams, and birds and snakes breathe the same air as you. Why *not* get to know them?

STELLER'S JAY

Watching wildlife is about finding animals, identifying them, and then seeing what they're up to. It doesn't take much. Just some curiosity, a bit of patience, and a willingness to learn. Wildlife watching can be as simple as noticing ants attacking cookie crumbs or as challenging as identifying nighttime animals by their shining eyes. It's up to you! There's always someone crawling, hopping, or flying by to learn about. Besides being super entertaining, watching the animals around you is an easy cure for boredom on car trips, waiting for buses, or walking to a friend's house. Check out something wild today!

Take Wild Notes

Writing down what you see will turbocharge your path to expert wildlife watcher. Documenting and reflecting on who and what you've seen helps you notice patterns and exercises your memory of different animals. Able to identify an animal to its species? Congrats! Check off its I SAW IT! box on the animal's page in the animal identification section, and write in some basic notes, too. Here are some more tips on note-taking:

DATE Use a full date, including year. Who knows when you might look back at your notes?

TIME If you don't know the actual time, *late afternoon* or *early morning* works, too.

LOCATION Be as specific as possible with location. *Juniper Park* is too general. Whereas, *Bench by Fifth Street entrance to Juniper Park, Springfield, Idaho,* is a location you (or someone else) can actually find and revisit.

WEATHER is whatever you know. You don't have to look at an app to know it's *chilly and raining* or *sunny and cold.*

NOTES is a place to jot down observations like, *sparrow pecked at ground* or *squirrel kept flicking tail.* It's also space for sketching what you saw or want to remember.

TRY IT → Watch Some Locals

Wild animals are all around. Need proof? See for yourself.

STEP 1 Go outside and sit for five minutes. Try to stay still and be quiet, blending into the background. The goal is for animals not to notice you or see you as a threat. Look around for birds, squirrels, or anything that's wild.

STEP 2 Check off any of these that you observed or noticed:

- ☐ animal flying
- ☐ animal making noise
- ☐ animal eating
- ☐ animal with wings

Anything else? ....

STEP 3 Now get up and walk around. Do you see any evidence that animals were there? Check off any you found:

- ☐ nest
- ☐ animal tracks
- ☐ snake skin
- ☐ holes in trees
- ☐ antler or bone
- ☐ scat (animal poop)
- ☐ fur or feather
- ☐ holes in ground

Anything else? ....

STEP 4 Fill out the TRACK IT on page 6 to record the details and make your first observations official.

I DID IT! DATE:

TRACK IT ↘ Wild Journal

There are TRACK IT sections throughout this book where you can answer specific questions, write down details, and record the critters you've spotted. Use your experiences and observations from the TRY IT sections to fill them out.

STEP 1 Fill out the information below based on the TRY IT activity on page 5.

DATE

TIME

LOCATION

WEATHER

What animal activity did you see?

What animal evidence did you find?

Can you tell who created the evidence?

Notes:

STEP 2 Could you identify any of the animals? ....

If so, find them in the animal identification sections of this book. (Look for the bands of color along the book's edge.) Got a positive ID? (Yay!) Check off the I SAW IT! box on its page and fill in the blanks. Yahoo! It's official!

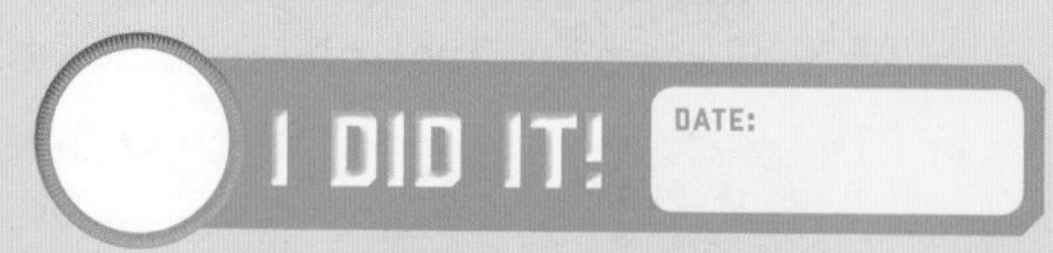

CHAPTER 2

Scout for Wildlife

Wild animals live nearly everywhere, but some places are better than others for spotting wildlife. Animals don't grocery shop or rent apartments. They have to find shelter and food in nature. The best sites to look for wildlife are natural areas that have what animals need to survive. Habitat that's good for wildlife is good for wildlife watching.

Knowing what wild animals need can help you find them. Wildlife needs water to drink, and food—either plants to eat or prey to hunt—as well as shelter that protects them from weather and predators. They also need space and supplies to breed and raise young. For birds that means nest-making places and materials, while for mountain lions it means dozens of square miles for males to patrol a breeding territory. Visit a field full of seedy plants in late summer and you're sure to see finches and other seed-snacking birds. If you're in a desert, wherever there's water is the place to be—for wildlife and their watchers.

Habitat that supports wildlife has food, water, shelter, as well as whatever's needed to raise young. Does this habitat have all four?

Remember that not all animals need the same kind of habitat. You'll find acorn-gathering chipmunks in a grove of oak trees with places to hide from hungry hawks. But that's not suitable habitat for all wildlife, of course. Looking for cutthroat trout? These cold-water fish need habitat with clear, gravel-bottomed streams so they can dig out nests to lay eggs. Toads need still water to lay eggs in but also bugs to feast on—good toad habitat has both. The value, or measure of goodness, of any particular habitat depends on who it's being measured for. What's best for elk isn't the same as what's decent for pelicans. How good a habitat is depends on who's trying to live there.

WOODHOUSE TOAD

Know Where (and When) You Are

This is why watching wildlife is endlessly fascinating. Who's around depends on where *you* are. Even hanging out in the backyard, playing soccer in a local park, or biking by a weedy lot can mean visiting different habitats. Each is a new puzzle to figure out who lives there. Every new place is a chance to notice birds, see squirrels, or spot something new. You'll hear different birds singing in a city park than along a forest stream. The fish in a lake aren't the same as those swimming in a creek.

When you're somewhere matters, too. Both season and time of day change the mix of active animals out and about. No reason to look for bats in the sky at noon, or chipmunks in the middle of the night. Pretty, migrating warblers and other songbirds aren't the only ones who flock to the Great Lakes in spring. So do birdwatchers! And autumn is the time to see bucks sporting their biggest antlers of the year. The more you learn about the animals you're looking for, the more likely you are to spot them. Enjoy the search!

CHIPMUNKS

Wildlife Spotting Tips

While you won't find hummingbirds and minnows in the same exact habitat, here are some general guidelines for spotting wildlife:

GO FOR THE EDGE(S) Any place where water meets land, grassy lawn meets bushes, or farm field meets forest is good for wildlife viewing. Why? Because these edges often have the best of both worlds for animals—open sunny areas where food plants grow, and trees or bushes for cover from predators.

WHITE TAILED DEER

AVOID MIDDLE-OF-THE-DAY DOLDRUMS The warmest, sunniest part of the day is naptime for many animals. Birdwatchers are early risers for a reason! Many furry animals are crepuscular, which means they're most active at dawn and dusk.

FAWN

SHUSH AND SIT Animals easily see you if you're standing in the open with the sky behind you. Wear drab clothing and try to be quiet, still, and low to the ground. When you do move, do so slowly and silently. Chasing animals doesn't work—and isn't nice—so let them come to you. Enjoy the small sounds and interesting smells as well as the sights as you wait!

Be Adventurous, Be Sensible

Watching animals shouldn't be itchy, ouchy, or scary. Here are some reminders to help keep it fun:

NEVER TOUCH FURRY WILD ANIMALS

Most furry animals have teeth and claws and they can carry the disease rabies. Treatment for a nip or scratch? Bunches of shots.

IN FACT, DON'T GET TOO CLOSE TO WILDLIFE If your presence makes an animal stop eating or run away, you're too close and are causing it stress. You're in their home; respect it.

STAY AWAY FROM WILD PARENTS Mama bears aren't the only animals protective of their young. Daddy geese are shin biters! Getting between (or too close to) animal parents and babies is never safe.

CANADA GOOSE

AVOID BITES FROM EIGHT-LEGGED CRITTERS

Leave spiders alone. Check for ticks and safely remove any you find. Get educated about the tick-transmitted illnesses where you are.

WATCH OUT FOR SIX-LEGGED STINGERS Don't swat at bees or yellow jackets, that makes them more likely to sting. If you're allergic to bees or wasps, carry whatever medicine you need.

DO SWAT MOSQUITOES Cover up at dawn and dusk to avoid bites. Find out if mosquitoes carry diseases where you are.

BE SNAKE SAVVY Find out if venomous snakes are around, and know what they look like. Check out page 325 for more tips.

LEAVES OF THREE, LEAVE IT BE Think you touched some poison ivy or oak? Wash your hands well as soon as possible, and use a poison ivy wash if you have it.

STAY OFF PRIVATE PROPERTY If you're not sure, don't enter without permission. Just because there's not a NO TRESPASSING sign doesn't mean it's okay.

LEAVE IT LIKE YOU FOUND IT If you move a log or rock or pick up a worm or toad, put it back.

TRY IT → Wild Things Come to Those That Wait

No one spots a rare snow leopard by casually glancing out the window. A wildlife watcher's ability to continue to look and listen, to keep returning their attention to the branches of trees and rustle of grass makes all the difference. Time is your friend. Spend enough time in one spot and something will fly, crawl, swim, or scamper by eventually. Still not a believer? Try it!

WHAT YOU'LL NEED

> a pencil or pen, a watch or clock

STEP 1 Find a comfortable spot outside. Write down the time now .... and in five minutes ....

STEP 2 During the next five minutes, jot down the names of any animals you see here:

STEP 3 Repeat steps 1–2 but for ten minutes.
Time now .... and in ten minutes ....
This time, write down the names of animals seen, and also keep a tally.

STEP 4 Repeat steps 1–2 for fifteen minutes. (Come on, you've got it in you!) Time now .... and in fifteen minutes .... Remember to write down the names of animals seen, and keep a tally if seen more than once.

STEP 5 Whew! Have a stretch, then look over the three lists.

→ Are there differences?

→ With more time, did you see more kinds of animals?

→ More total numbers of animals?

→ Did more time enable you to notice more detail, or discover something surprising?

TRACK IT ↘ Wildlife Habitat Survey

WHAT YOU'LL NEED

➢ a pencil or pen

STEP 1 Go out into the backyard, to a park, or someplace with a view of nature.

STEP 2 Sit in a single spot and draw the view you see. Include trees and bushes, grass and puddles—whatever is in your view.

DATE

TIME

LOCATION

WEATHER

My View

STEP 3 Think about what birds, fish, reptiles, amphibians, and mammals need to survive. Is it here? Do the trees make nuts? Are the bushes safe for nests? Is something tasty living under that log? Is there water? Remember the four habitat needs of wildlife (water, food, shelter, young-raising resources)? Check off those you see, and write the source. For example: a pond would be a water source, acorns a source of food, bushes are shelter, and a hollow log a place for a den.

- [ ] water:
- [ ] food:
- [ ] shelter:
- [ ] resources for raising young:

STEP 4 Walk around the area you've just drawn. Turn over rocks and logs, look under bushes, and check out the trees.

→ Did you see any animals or evidence of them (prints, scat, nests, burrows, etc.)?

→ Who might this place be a good habitat for (birds, fish, frogs, squirrels, etc.)?

CHAPTER 3

Become an Active Observer

Wildlife watching is a skill. Being able to scan an area, notice signs of animals, and identify who is flying or hopping around takes practice. Like most skills, you'll get better with time. Learning to recognize the wild animals around you is so worth it. You'll see things you never knew were right there all along. Discovering a new world takes time to understand. Plus, practicing is easy, cheap, and fun! Who doesn't want to spend time outside looking for critters? It's always an adventure.

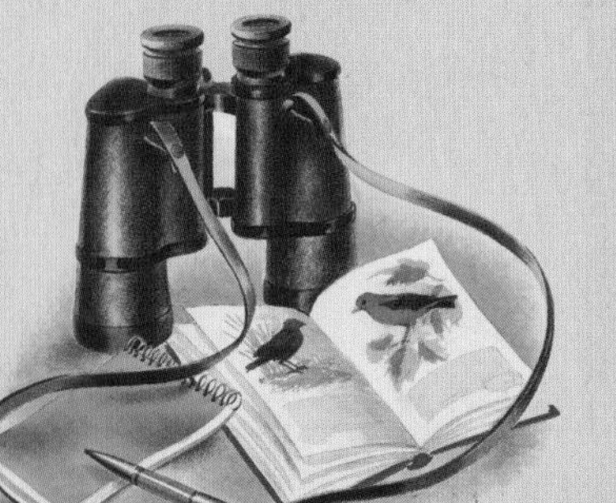

Never underestimate the power of paying attention. Observing is how professional biologists study animals. They make historic discoveries by watching with their very own eyes—and you can, too. Of course, scientists also document their observations. Writing down and drawing or photographing what you see takes wildlife watching to another level. Taking notes or photos helps you follow up on any questions you have as you're observing and look something up later. Drawing can also focus what you see and help you notice details—and remember them.

Essential Tools

Two essential tools of the wildlife watching trade are binoculars and a field guide. Binoculars, or binocs, let you see up close without being too close. Through binocs you can see whiskers and scales without scaring away or bothering animals. Binoc sizes are a two-number code separated by an X, like 7x35 or 12x50 (say: twelve by fifty). The first number is magnification. Binocs that are 10x make things looks ten times bigger, for example.

You'd think the more magnification the better, but zooming in shrinks the field of view—what you can see. It's hard to scan for birds in a tree if your binocs are so powerful that a bird's foot takes up the whole field of view. A magnification of 7x or 8x is pretty good for watching animals. The second number is the diameter (in millimeters) of the big front lens. Bigger lenses let in more light, so they are better for seeing in dark forests or at dusk. But the tradeoff is having to carry all that heavy (and expensive!) lens glass.

Want to know what it was you saw so clearly through the binocs? That's what the identification sections of this book are for. Each animal identification (or ID) section features commonly spotted birds (see page 86), mammals (see page 240), amphibians (see page 338), reptiles (see page 358), or fish (see page 406). Not finding what you saw in these pages? A field guide is your next step. A field guide is a book or app meant to be used outdoors (in the field) to identify animals. They often focus on regions and specific kinds of animals—they can be as specific as Beetles of Texas.

One of the most useful parts of any field guide is its range maps. (This includes the range maps in the identification sections of this book.) It should be one of the first things to check when identifying an animal. If an animal doesn't live where you saw it, chances are that's not it. Some range maps are simple, like this one:

The orange color shows where the American beaver lives.

Other kinds of range maps are more complicated, especially for birds that migrate. The maps tell you not only *where* the birds live, but *when* during the year they live there. Carefully read the explanation of the key, its colors and patterns, and term definitions in the front of the field guide first.

This American goldfinch range map shows Canada as only a summer (breeding) home.

Not all range maps are that complex, so don't worry. Just remember when reading North American keys that *summer = breeding* and *winter = nonbreeding*. That will tell you that if you saw a bird with deep blue wings in Illinois in January, it wasn't likely an indigo bunting. Try again!

Gear Up!

In addition to binoculars and the book you're reading right now, this stuff will come in handy for wildlife watching. Grab it on your way out the door:

Essential Skills

There are specific tips for spotting and identifying birds, mammals, and fish. But here are some general guidelines for noticing and identifying wildlife.

MOCKINGBIRD

PAY ATTENTION Sounds obvious, but we've all walked through a room and not noticed who's there. Think of it as being a wildlife detective. Look for clues, take note of who's around, and notice things that are out of place.

PATIENCE IS A TOOL Use it. The more time you spend in a place, the more you'll see. If a fox has been seen around a certain creek in the evening, you can't go once for five minutes and expect to see it. Patience . . .

LOOK ALL AROUND YOU Check out what's flying overhead and look for movement in the trees. Stare at the bushes and the ground beneath them. Search underneath logs and rocks. Look left and right, and turn around! Something might be behind you.

USE ALL YOUR SENSES Wildlife watching is also wildlife listening and wildlife smelling. Animals make sounds as well as noises when they move. *Plop! Plop!* Could be frogs jumping into the pond. Nothing says *Bats live here!* like the musky smell of guano, or bat poop.

THINK WHILE OBSERVING When you see an animal, think about what it looks like and what it's doing. Describe it to yourself: *Small round bird, dark stripes on its head, hopping from plant to plant eating seeds.* This will help you notice and remember its details.

LOOK IT UP Start with the identification sections in this book, and then perhaps invest in a field guide that is focused on your geographic location or animal of choice. Libraries have field guides, too.

Wildlife Identification's Fantastic Five: Shape, Size, Color, Behavior, Location

Whether you're birding, studying squirrels, trying to tell toads apart, or determining sunfish species, the core clues to look for and consider when making an ID are the same. Learn them and train yourself to notice them, and you'll be on your way to separating finches from phoebes, chipmunks from ground squirrels, and perch from puffers.

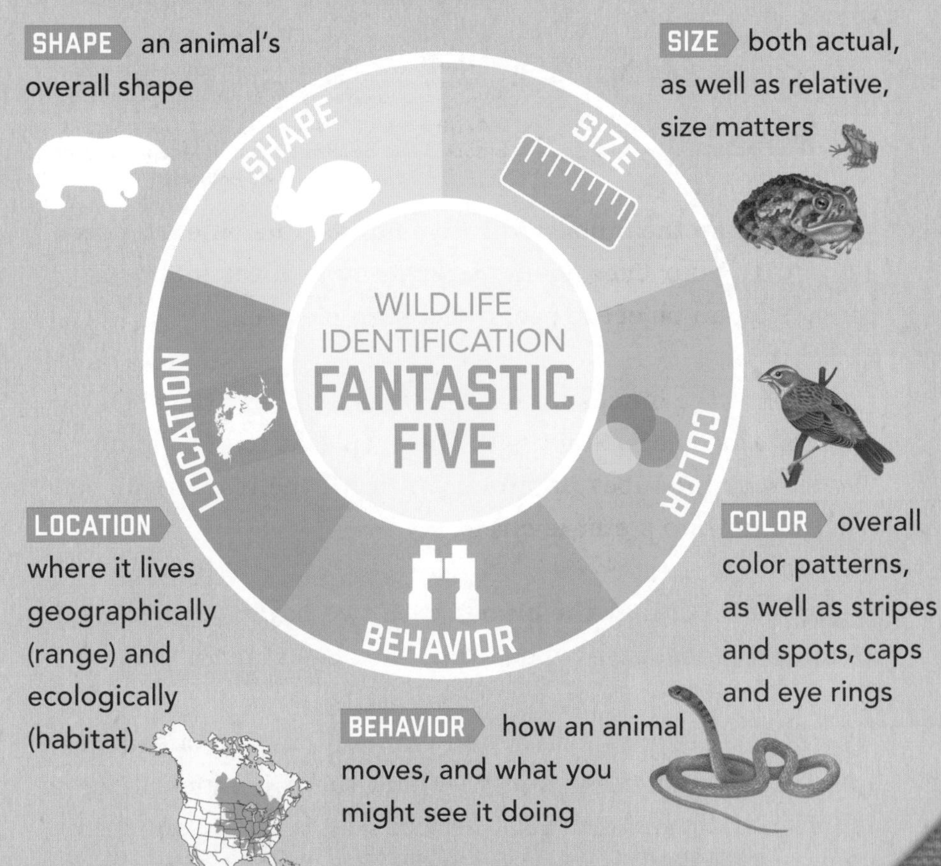

SHAPE an animal's overall shape

SIZE both actual, as well as relative, size matters

COLOR overall color patterns, as well as stripes and spots, caps and eye rings

BEHAVIOR how an animal moves, and what you might see it doing

LOCATION where it lives geographically (range) and ecologically (habitat)

TRY IT → Focusing In

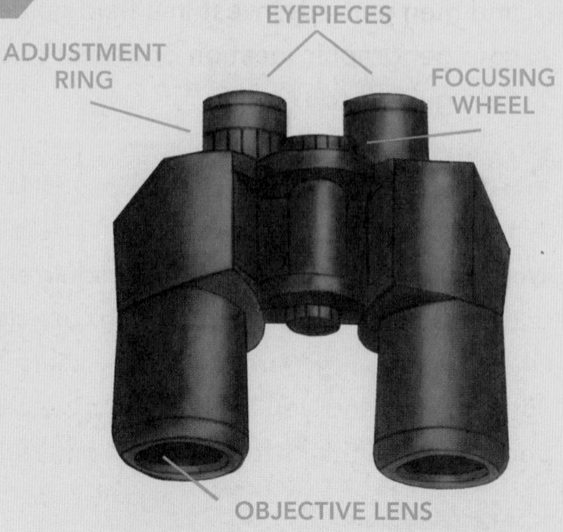

Focusing binocs takes some practice. The key is to first look at your target animal or object, hold your head still, and then bring the binocs up to your eyes. That makes it easier to find the object once you're zoomed in.

WHAT YOU'LL NEED

- binoculars

Note: Not all binocs are the same. If you're having trouble focusing in on what you want to see, find the user's manual that came with the binocs. Long gone? Look up the manufacturer's website and download it.

STEP 1 Grasp the binocs with two hands, one on each tube. Look with your eyes about as far away as a school bus is long. Pick an object to see, like a sign or tree.

STEP 2 Put the binocs up to your eyes and look through them at the object. Start with the eyepieces too far apart. Then move the tubes up and down until the two circles you see line up into a single circle.

STEP 3 Still holding the binocs with two hands, use a finger to move the focusing wheel until the object isn't blurry.

STEP 4 If the binocs have an adjustment ring on one or more eye, focus those. Close the eye where the adjustment ring is (often the right) and use the focusing wheel to sharpen the object for the open eye. Then close the opposite eye (the side with the adjustment ring) and use the adjustment ring to focus that eye. This adjustment ring setting shouldn't change the next time you use your binocs. It's for your personal eyesight, not what you're looking at.

TAKE IT TO THE NEXT LEVEL ↗

Range Map Search

The first step of wildlife watching is knowing what animals live where you're looking. If you're in Maine, don't go looking for roadrunners—none live there. How do you find that out? You look at a range map for greater roadrunner, find the state of Maine on the map by its shape, and see if it's shaded in. (It isn't.) Ready to get some practice?

STEP 1 Browse the identification sections. They show up as bands of color on the book's edge. Each color is for a different animal group. Pay attention to the range maps. Can you find your state on them? Do you know where in the state you live?

STEP 2 Choose one local animal in each section that lives where you do. Write it in below.

Bird:	Mammal:
Fish:	Amphibian or Reptile:

STEP 3 Try to find that animal in the wild. Mission accomplished? Woo-hoo! Check off the I SAW IT! box on the animal's page and fill in notes, too.

PART II

BIRDS

Late spring is your favorite time of year. Especially when it comes to bird-watching! Winter is over, and all the migrating birds have come back and are wearing their brightest breeding colors. You're out for a walk, binoculars at the ready. Birds are frantically flying back and forth to their nests with food for hungry baby birds. Everyone's busy! You're watching a robin's nest through your binoculars. Two little featherless baby birds peep with huge open yellow triangle beaks. So cute! Then you notice the peeping coming from down below. There's a baby bird on the ground! It must have fallen out. Should you try to put it back in the nest? Or will your scent make the parents abandon all the hatchlings? *What would you do?*

CHAPTER 1

How to Be a Birder

RUFOUS HUMMINGBIRD

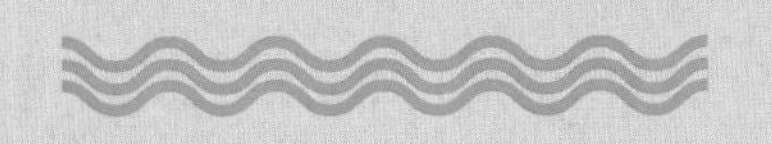

INDIGO BUNTING

If you want to watch wildlife wherever you are, birds are your go-to critter. Just look up! Birds are captivating creatures flying around in colorful feathery outfits. Birds live in neighborhood parks and backyards, in cities and towns, on farms and in fields, at the beach and in the mountains, in the desert and the Antarctic. (And who doesn't love penguins?) There are thousands of kinds of birds on planet earth. North America alone has more than seven hundred different species of birds. Lifelong bird-watching enthusiasts, or birders, never run out of birds to see or places to find them. And birds are endlessly interesting to watch. They're active daytime animals that make all kinds of sounds, come in many sizes and colors, and often interact with one another. And birds do something extra special—they fly! Seeing a pair of hawks soaring high overhead never gets old. A noisy, flapping flock of glossy blackbirds doesn't go unnoticed! And hearing the wings of a hummingbird is always exciting.

SHARP-SHINNED HAWK

Birds are the perfect subjects for beginner wildlife watchers. Why? First off, most birds are diurnal, which means they're out and about during the day. That makes them much easier to find than nocturnal animals, like bobcats and foxes. And unlike frogs and snakes, rabbits or chipmunks, birds aren't usually hiding. Many animals are secretive and difficult to spot. You could walk right by them in the woods and not even know they were there. Camouflaging and concealing themselves is how they survive. By comparison, birds are loud, rowdy, colorful, attention grabbers! That makes them easy to find and fun to watch. Bird-watching is simple to start and never gets boring. There's always more to see and new ways to level up your birder game. Ready to start your life list?

Talk Like a Bird-watcher

Here are a few terms to know:

BIRDER someone who likes to go out bird-watching; a bird-watcher

FIELD MARK distinctive feature useful in identification

LIFE LIST a birder's personal list of every species seen

LIFER a first-time-ever bird species sighting for a birder

LBJ (LITTLE BROWN JOB) a drab brownish songbird that looks like a lot of other drab brownish songbirds

NEMESIS BIRD a bird species that a birder has repeatedly failed to see

ORNITHOLOGIST a scientist who studies birds

TWITCHER a seriously hard-core birder who is all about racking up lifers

VAGRANT a bird outside its normal ecological or geographical range

FIELD SPARROW

TREE SPARROW

SONG SPARROW

LBJs can be hard to tell apart!

See Who's Flying Nearby

Birds are everywhere and are easy to see—just go out and look.

BARN SWALLOW

WHAT YOU'LL NEED

- a pencil or pen

STEP 1 Go outside and sit quietly for ten minutes.

STEP 2 Check off any that you observed or noticed:

- ☐ white bird droppings
- ☐ bird perching
- ☐ bird flying in air
- ☐ group of birds
- ☐ bird sounds (songs, calls, or wingbeats)
- ☐ bird nest or feathers

STEP 3 Did you recognize any of the birds you saw? Do you know their names? Flip to page 88 and scan the Common Neighborhood Birds in the Bird Identification section. Make any positive IDs? Nice! Check off the I SAW IT! box and fill in the details.

CHAPTER 2

Natural Born Flyers

Want to spot awesome birds? Learn a bit about what makes birds unlike any other kind of animal—feathers! Knowing more about feathers will boost your bird IDing chances. Birds are the only animals with feathers. And while feathers are for flying, even birds that cannot fly (kiwis, penguins, ostriches, etc.) are covered in feathers.

Fabulous Feathers

Feathers cover the wings of birds, as well as other body parts. Wing feathers create a lightweight surface for catching and pushing air around during flight. Feathers flex and bend but don't break, thanks to keratin. It's the same super-useful, tough, flexible, fibrous protein that's in hair, claws, and horns. Like hair, feathers grow out of the skin. Besides giving wings a flight-ready covering, feathers are excellent insulation. They keep birds warm and shed water like a rain slicker coated in body oil.

Even flightless birds, like this penguin, are feathered.

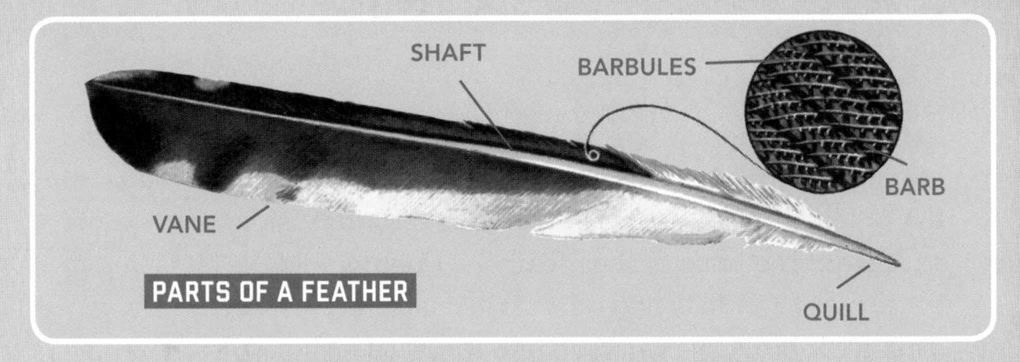

PARTS OF A FEATHER

Feathers make the bird. Feathers are pretty amazing products of evolution. Different birds have different kinds of feathers, but most have a mix of body and flight feathers.

- Fluffy down feathers often grow next to the skin. They're mostly just fluff.
- Contour feathers are small body-covering feathers that lie in layers like roof shingles. They're often rounded.
- Tail feathers are for steering. They are usually long, with a center shaft and equal-size feathery sides, called vanes.
- Wing or flight feathers have long center shafts but uneven vanes—one side is thicker than the other. The side that leans into the wind is narrower.

Got a Permit?

Leave any feathers where you found them. Collecting the feathers of native North American wild birds without a permit is illegal in the United States. Why? Many species of birds nearly went extinct because of the value of their feathers. Hunters sold the birds' feathers to makers of hats and other fashionable accessories. That's why most birds are now protected and illegal to hunt, kill, harm, or collect. This includes feathers because there's no way to be sure the bird wasn't killed or harmed to get them. But observing, drawing, and photographing feathers is okay—as long as you leave them in the wild afterward.

Built for Flight

Flying is an astonishing adaptation. Being able to quickly and easily travel from one place to another gives birds huge advantages in nature. Barriers like highways and flooded rivers stop some animals from finding food or mates. Birds have the ability to search farther for what they need because of flight. Hundreds of species of birds migrate across continents (and back!) every year to be in the best places to find food or raise young. That's not something a snake or a chipmunk can do.

Hollow bones (white) and large lungs (blue) are used to pump oxygenated blood through the heart (light red) to other strong muscles (dark red) and power bird flight.

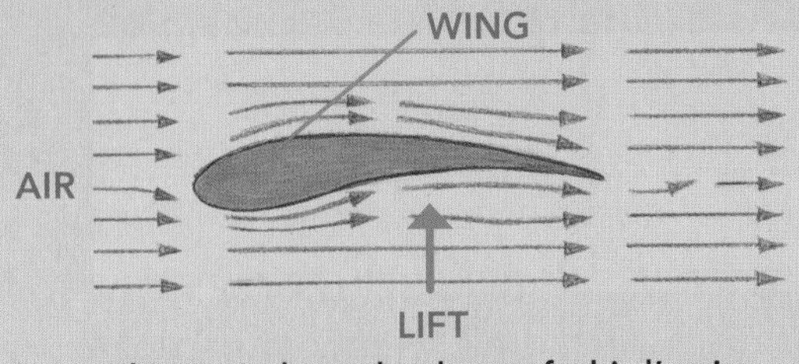

Like an airplane, the shape of a bird's wing creates lift.

Flight adds to the fun of bird-watching, too. Birds migrate through areas, so the mix of birds to spot changes throughout the year.

Birds have a pair of legs and two wings for front limbs. Large muscles power the flapping of wings. They attach to a bird's oversize breastbone. The entire body of a bird is built to fly—both inside and out. Flying is about being able to move forward fast enough for air to flow over a wing and create the force, called lift, needed to hold something in the air. And weight is the enemy of lift. Lightweight things get up into the air easier. Birds are vertebrates, animals with backbones like leopards and lizards. But the skeleton of a bird is lightweight for its size. Bird bones aren't very dense and many are hollow. This cuts down on weight. Think about how fragile a chicken leg is compared to a ham bone. Now you know why the saying "when pigs fly" means never! Bird bodies are also streamlined. Their aerodynamic shape reduces air friction, which boosts flight.

CRITTER CONFUSION: Bird or Bat?

BATS

BIRDS

Don't confuse bats with birds like swifts or nighthawks. These evening insect-eating birds have wings that point back behind them, like an airplane. Bats flutter their upturned wings.

TRY IT → Sorting Out Feathers

Birds are constantly losing and growing, or molting, feathers. Hunt for feathers in a park, backyard, or nearby trail. Feathers can carry germs, so be sure to wash your hands thoroughly after handling them. Or wear disposable or dishwashing gloves to pick them up.

WHAT YOU'LL NEED

- gloves or moist towelettes, a hand lens or magnifying glass (optional)

STEP 1 Look for feathers wherever you see birds. Near feeders, at the base of trees favored by birds, near fence posts, etc.

STEP 2 When you spot a feather, take a good look. Which kind of feather do you think it is? Use the descriptions and example pictures of the four kinds of feathers on page 31 to help you decide.

Check off which kind it is:

- ☐ down
- ☐ contour
- ☐ tail
- ☐ wing feather

STEP 3 Use the hand lens to get a closer look at the barbs branching off the shaft. The tiny bits branching off the barbs are barbules. Try pulling them apart and re-zipping them back together.

STEP 4 Wash your hands if you touched anything!

TRACK IT ↘ Draw a Feather

Want to remember what the feather you identified looks like? Draw it! It's a terrific way to notice details and build memory.

WHAT YOU'LL NEED

- ➢ a pencil or pen, colored pencils or pens

STEP 1 First draw the shaft, paying attention to its length and how much it narrows toward the tip.

STEP 2 Notice how even, or uneven, the barbs are on both sides of the shaft. Draw them.

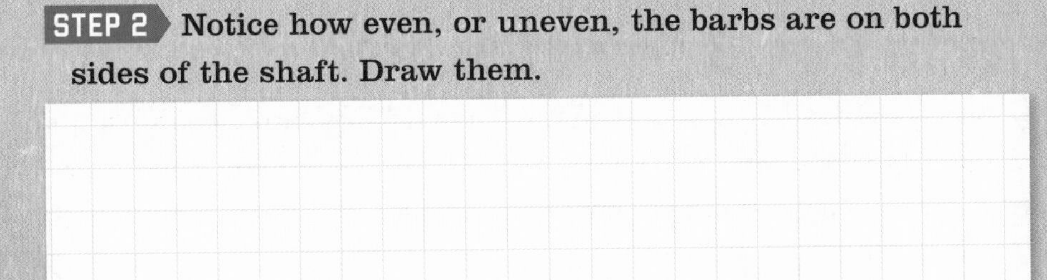

STEP 3 Add in any other details you see, including colors.

→ Label what sort of feather (wing, tail, contour, or down) you think it is.

→ Any guesses on who lost it? (Hint: Look around, maybe it's still nearby!)

CHAPTER 3

Spot Birds by Silhouette and Shape

Part of the fun of bird-watching is being able to identify the birds you see. Identification is important for a number of reasons. Remembering which bird is which is easier when you know their names—and knowing names helps you remember what birds look like. Plus, it's difficult to share what you've seen with others when you don't know bird names. Not knowing names gets problematic quick.

So, how should you go about learning to identify birds? Flipping through the Bird Identification section can be a bit overwhelming—there are so many birds! But first, give yourself some credit. You know more than you think. Can you tell a hummingbird from an owl? A goose from a hawk? Of course, you can. Learning to identify birds is about seeing shapes and sizes, colors and patterns. Bird identification is also about knowing where you are and paying attention to what the bird is doing.

Shape? Size? Color? Behavior? Location?

WILDLIFE IDENTIFICATION
FANTASTIC FIVE

SHAPE
SIZE
COLOR
BEHAVIOR
LOCATION

Let's break it down. Most birders use five clues to identify a bird.

We'll start with shape. Overall shape goes a long way when identifying birds. Ducks have a duck shape, after all. Is the bird leggy and long necked, like a heron or egret? Pointy-winged with a forked tail, like swallows? A bird's overall shape helps to pinpoint its group. Once you've recognized a bird's group, the identification possibilities shrink. Going from *What kind of bird is that?* to *What kind of owl is that?* is a huge step.

Describing birds by their silhouette shape is common in field guides. Silhouettes are a great tool for learning bird shapes. Why? Part of what's great about birds is their colorfulness. But that's distracting when you just want to learn its shape. Think about the difference between a brightly colored male bird and its mate. So different, right? But their silhouettes are exactly the same. The single color of a silhouette reduces a bird to just its basic shape.

How many of these birds can you recognize just by their shape?

TRY IT → Spotting Silhouettes

With hundreds of birds out there, identifying the one in front of you starts with its overall shape—its silhouette.

WHAT YOU'LL NEED

- a pencil or pen

STEP 1 Look at the eight bird silhouette shapes. Read the labeled groups they represent. How many are you familiar with? Probably most, if not all.

STEP 2 Start bird-watching! Go for a walk, sit in a park, or just go about your day. Pay attention to the birds around you. Which groups did you see?

STEP 3 Check off the bird groups as you see those bird types. Fill in the date, place, and any observations, too.

HUMMINGBIRD

When I saw it:

Where I saw it:

Observations:

When I saw it:

Where I saw it:

Observations:

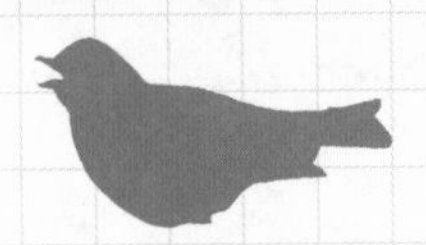

FINCH

When I saw it:

Where I saw it:

Observations:

OWL

When I saw it:

Where I saw it:

Observations:

DUCK

When I saw it:

Where I saw it:

Observations:

DOVE

When I saw it:

Where I saw it:

Observations:

WOODPECKER

When I saw it:

Where I saw it:

Observations:

HAWK

When I saw it:

Where I saw it:

Observations:

TRACK IT ↘ Be a Bird Artist

Drawing is a terrific birder skill. It trains the eye to observe and the brain to remember! The shapes of bird groups is the perfect place to start. You can draw nearly any bird shape with an oval for the body and a circle for the head. Here's proof.

See how the body ovals and head circles are simply arranged differently? The gull's body oval is horizontal while the owl's is upright, or vertical. The owl's head circle is bigger compared to its body than the gull's, too. And neither has much neck showing! Notice how the wings, tails, beaks, necks, legs, crests, and other parts are easier to add in after the head circle and body oval are properly placed. Grab some paper, practice by tracing first, then go for the real thing.

WHAT YOU'LL NEED

- a pencil or pen, thin paper that's easy to see through

STEP 1 Place the paper over silhouettes on pages 38-39.

STEP 2 Draw ovals over the bodies and circles over the heads.

STEP 3 Add a neck, legs, a beak, and other details to finish each bird's picture.

STEP 4 Ready to go freehand? Choose one of the silhouettes as a model. In the space below, first draw the body oval and the head circle. Then fill in the details. You're a bird artist!

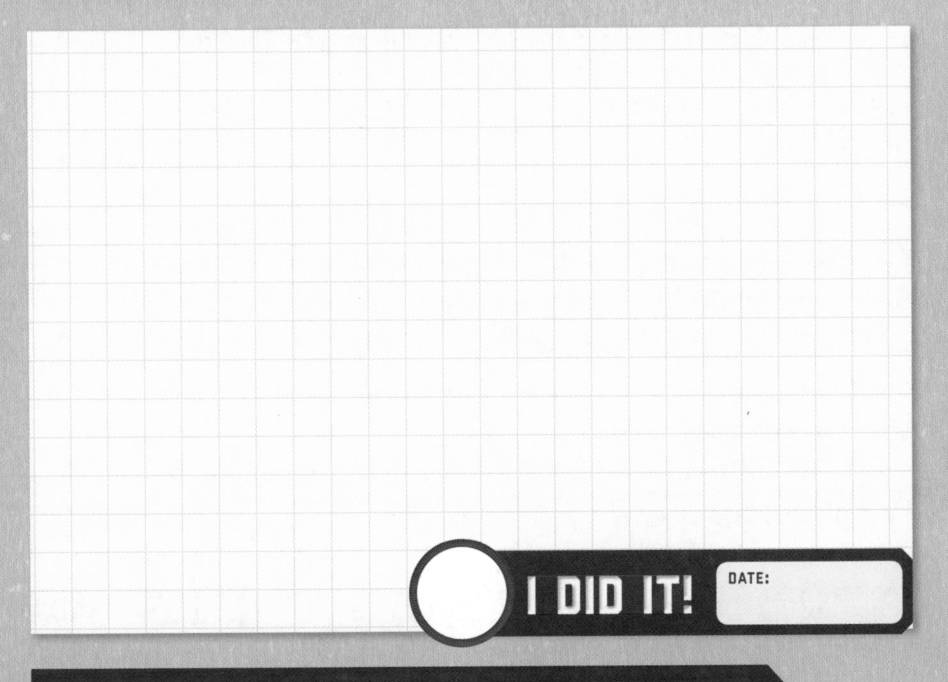

TAKE IT TO THE NEXT LEVEL ↗

Recognize Shapes in a Flash

Are these bird groups and shapes unfamiliar to you? Turn them into a memory game. Simply draw the shapes onto index cards or slips of paper, and write the group names on the other side. (Just by making your deck of cards, you'll have them memorized in a flash!) Make an additional set of cards on different colored paper (or use a different colored pen). Then lay out the one set of cards names up, and the other pictures up. Test your own skills or play with a friend—whoever finds the most matching sets wins!

CHAPTER 4

Scout Birds by Size and Place

Size is an important clue to identifying birds. Think about it! A big sparrow is tiny compared to a small swan. No need to flip through field guide pages of finches when you see a bird as big as a hawk. Of course, words like *big* and *small* are relative. And while measurements are great, most birds aren't going to hold still while you find a measuring tape.

When out bird-watching, birders often use familiar species as a kind of bird-size unit.

A **SPARROW** is about 6 inches (15 cm) beak to tail tip, about an inch longer than a soda can on its side.

A **ROBIN** is about 8½ inches (21½ cm), a little bit longer than a brick.

A **CROW** is about 17 inches (43 cm), about as long as a clothes hanger is wide.

Birds are described as the size of a sparrow, robin, or crow. As in, *A robin-size bird with a blue head,* or, *It was smaller than a sparrow.* (Gotta know the lingo when hanging with birders.)

Habitat Matters

Habitat is another bird recognition clue. Knowing which birds belong where is a huge help when it comes to identification. If a bird is on a desert cactus, it's not likely a water-loving kingfisher. Spot an orangish bird on your feeder in San Francisco? Probably not a Baltimore oriole. And while there are more than a dozen kinds of hawks in North America, not all of them live everywhere all year round. (This is what range maps are for.)

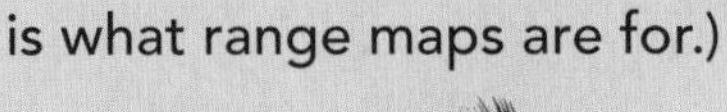

Belted kingfishers (left) don't hang out near cacti. But Greater roadrunners (right) do.

When and where *you* are matters, both geographically and ecologically. The mix of birds changes from marsh to woodland, seashore to open ocean, mountaintop to valley. And from season to season. Just because a range map says a bird lives in your state doesn't mean it's found in every nook and cranny there. Range maps show a bird's geographical range—the counties and states where the bird lives. But a shaded-in area on a range map covers lots of different places—cities and towns, farms and forests. Some of those habitats are suitable to sandpipers, while others are good for goldfinches. Brown pelicans live in California, for example. But they eat ocean fish, so you won't find them flapping around in California mountains or deserts. Birds have an ecological range within their geographical range.

Spot Bird Units in Action

Get familiar with sparrows, robins, and crows so you can use the sizes of these three birds in nature. Think of these birds as living rulers or yardsticks!

WHAT YOU'LL NEED

- binoculars (optional)

STEP 1 Review the sparrow, robin, crow silhouettes on page 42. Check out their pictures on pages 90, 93, and 107.

STEP 2 Go outside and look for the three measuring birds. When you see one, look around as it hops on the ground or lands on a park bench. What is its size compared to other things you know?

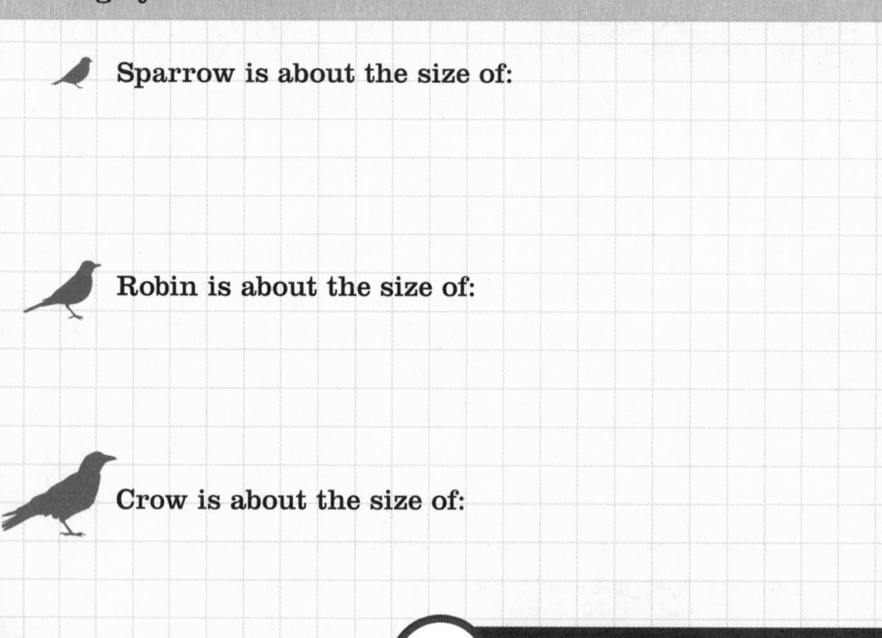

TRACK IT ↘ Size Scavenger Hunt

Put the measuring birds to good use. Scout a new bird, try to name what you saw, and draw a sketch of it. Can you find:

- [ ] a bird bigger than a crow
- [ ] a bird smaller than a sparrow
- [ ] a bird sized between a robin and a crow

TAKE IT TO THE NEXT LEVEL

Hone In on Habitat

Do you know your state bird? Do you know if it lives where you do in your state? If it doesn't, what's a common local bird? Use the information in the Bird Identification section (pages 86–187), including the range maps, to come up with birds that live:

- **In your state and close to your home.**
- **Far from your home but near a place you regularly visit.**
- **In a part of your state that is dry.**
- **In an area of your state near the shore, or a lake or river.**

Have a few candidates for your next bird-watch? Now go out and find 'em!

CHAPTER 5

ID Birds by Color

Birds come in a gorgeous rainbow of colors. Their feathers have patterns, spots, and stripes in every hue under the sun. The colors of birds are part of their appeal. Colors are also key clues to what bird it is! Even a fleeting glimpse of an all-red bird narrows down its identification, because there just aren't that many. (Probably a male cardinal, possibly a summer tanager.)

Team All Reds! Northern cardinal (left) and summer tanager (right).

When you're flipping through a bird field guide, colors pop out at you. But color can also lead you to the wrong bird if you're not careful. Color is variable. Think of all the different kinds and shades of blue there are, for example.

Who's the bluest bird of all?

Birds are often spotted among leafy branches or

against the bright sky, so sometimes seeing a bird's color isn't as easy as you might think. Feather color can vary within a species, too. Age, sex, health, and season can all affect color. So birders use overall color patterns rather than absolute colors to help sort out birds. Was it red-chested with a dark head? Mostly yellow with black wings? All white with yellow legs? These kinds of quick-and-easy color descriptions are super useful for identifying birds—especially when combined with a size (*sparrow-size with red chest and dark head*) and/or a group (*sparrow-size woodpecker with red chest and dark head*). Bird size, color, and shape all add up to narrow down possibilities.

Birder Pro Tip

Good light is needed to see color well. (Ever notice how that blue shirt looks gray in dim light?) But bright sun at the wrong angle gets in the way of birding by color. If you're looking at a bird with a bright sky in the background, the bird will often show up as a silhouette. Back lighting (as photographers call it) is great for noticing a bird's shape, but you can't see colors well on a silhouetted bird. Try changing to a darker background to get a more colorful look. How? Move to shift the angle between yourself and the bird so the sun isn't directly behind it. Choose a line of sight that places a building, tree trunk, or something else behind the bird. Then its colors will be easier to see.

SUNNY BACKGROUND

DARK BACKGROUND

Color Differences Among Sexes, Ages, and Seasons

FEMALE AND MALE NORTHERN CARDINALS

You might know (or not, no biggie) that both the red birds on page 47 are male. Neither female cardinals nor summer tanagers are red. When males and females of a species look different, it's called sexual dimorphism. (*Dimorphism* means two [di] forms [morph].) Many male birds are more brightly colored than the females. Being showy helps males attract a mate. Sometimes the duller colors of a female help her hide on the nest, too.

Younger birds also sometimes look different. This isn't the difference between a newborn and its parent. That's a given. We're talking about immatures or juveniles, full-size birds that are as big as adults but aren't yet old enough to breed. For example, the heads of immature bald eagles are brown. They don't turn white until around age five.

BALD EAGLES

GOLDFINCH (MALE)

Time of year can matter, too, when it comes to color. Some male birds wear winter feathers, or plumage, that's distinct from summer plumage. Male goldfinches lose their bright yellow jackets in the fall and spend winters in duller plumage.

Colors That Make a Mark

Hey, check out that bird . . . whoops, it's gone. Birds are busy creatures that fly away in an instant. It can make getting a good long look difficult. That's part of the challenge! When you see a bird, concentrate first on its size and shape. If it's a bird you still don't recognize, look at the colors on its wings and head. These two body parts often have field marks, particular characteristics or features that clue you into an animal's identity.

WINGS Is there a colored *wing patch*? Are there lines of color, called *wing bars*, going across the wings? Are they dark or light? How many?

HEAD Look for a colored ring of skin or feathers circling the eye, called an *eye ring*. A line of colored feathers that goes through the eye is an *eyeline*. *Eyebrow stripes* run above the eye.

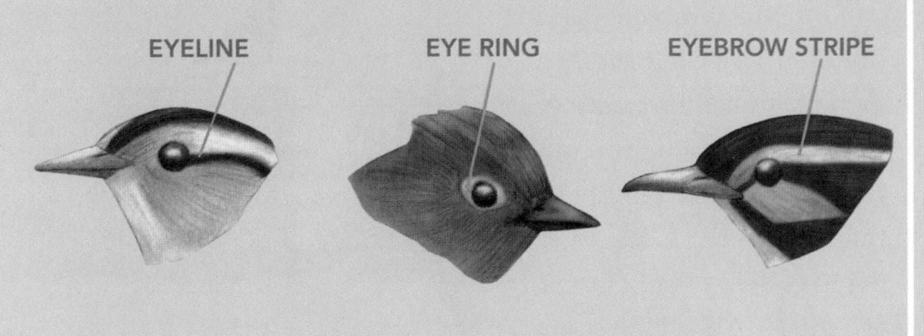

TRY IT → Spotting Color Patterns

Birders learn to notice a bird's overall main color patterns, not just one particular color or colored part. For example: Rather than try to spot an American robin by its red-orange breast alone, a birder looks for a dark head, gray back, and red-orange underneath. Got it? Then you're ready!

WHAT YOU'LL NEED

- a pencil or pen, binoculars (optional)

STEP 1 Go find some birds! Try perching on a park bench, taking a walk through a field, sitting near a feeder, hanging out at the beach, or going wherever birds are.

STEP 2 Choose a bird to focus on, call it Bird #1. Notice its overall color pattern—not just the one bright color that stands out. Look at the bird from top to bottom (head to feet) and side to side (wings and tail).

STEP 3 Answer the questions on the following page.

STEP 4 Did you recognize the bird you observed? Do you know its name? Nice! Look for it in the Bird Identification section (pages 86–187). Make a positive ID? Nice! Check off the I SAW IT! boxes and fill in the blanks.

Bird #1

DATE

TIME

LOCATION

WEATHER

Describe the bird's color pattern:

Any field marks on the wings or head? Check off which ones and write in the color.

☐ wing patches: ....
☐ wing bars: ....
☐ eye rings: ....
☐ eyeline: ....
☐ eyebrow stripes: ....

Bird #2

DATE

TIME

LOCATION

WEATHER

Describe the bird's color pattern:

Any field marks on the wings or head? Check off which ones and write in the color.

☐ wing patches: ....
☐ wing bars: ....
☐ eye rings: ....
☐ eyeline: ....
☐ eyebrow stripes: ....

I DID IT! DATE:

TRACK IT ↘

Colors to Remember

Draw each of the two birds from the previous TRY IT activity. Start with the body oval and head circle method from page 52. Use colored pens, pencils, or markers to shade in the colors you saw. Pay special attention to wing and head field marks! They're bird identification gold.

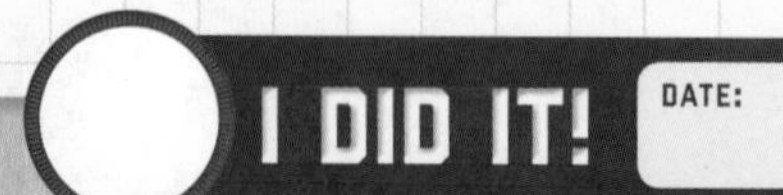

TAKE IT TO THE NEXT LEVEL ↗

Field Mark Hunt

Spot as many of these colorful characteristics of identification as you can. You don't have to spot them all in a day. Just keep looking whenever you're outside and include the date and place below.

- [ ] red on its head
  DATE
  LOCATION
- [ ] a yellow beak
  DATE
  LOCATION
- [ ] stripes or bars on its wings
  DATE
  LOCATION
- [ ] a white throat
  DATE
  LOCATION
- [ ] black eyebrow stripe
  DATE
  LOCATION

I DID IT! DATE:

CHAPTER 6

What Birds Do

Observing what birds are up to is the entertaining part of bird-watching. Behaviors are also terrific tips for identifying birds. Quail don't dive around in streams. Wild turkeys don't soar high overhead. The clingy, upside-down way nuthatches hop on tree trunks gives them away. And no other bird has the helicopter flight moves of a hummingbird. Birds are always on the move—flying, looking for food, singing, and more flying! Bird behavior is one of the five identification clues and probably the most fun one to look for. Go to any beach and people will be looking at terns and gulls plunging for fish and shorebirds skittering across sand. Bird feeders and birdbaths are popular because it's fun to see snacking sparrows or bathing buntings. Birds aren't performing for us, of course. They're just going about the business of survival. All bird behaviors serve a purpose—helping birds eat, get around, or stay safe. Let's explore some of those you might see or hear while outside.

Nuthatches seem to defy gravity.

The looping courtship flight of ruby-throated hummingbirds.

Protection from Predators

Many birds are social, banding together in flocks to run off predators. Safety in numbers! Ever see a group of crows or jays flying and diving at a hawk to chase it away? It's called mobbing and is quite an air show! Mobbing often means a nest is nearby. Birds also protect themselves by hiding. Many female birds are feathered in camouflage colors to blend in with the nest.

Crows mobbing a great horned owl. Get out of here!

Grooming

Feathers are high-maintenance outerwear. Most birds have a gland where the tail starts that makes preen oil. When you see a bird reaching its beak under its tail and then pulling feathers through its beak, the bird is preening. The oil moisturizes the feathers, keeping them from becoming brittle and breaking during flight. Pulling the feathers through the beak aligns them into a smooth surface with no gaps. This makes for a warmer, more waterproof covering and a better surface for flying. Preening also cuts down on irritating body lice and feather mites. When feathers get overly oily, a good dust scrub is called for. You've probably seen sparrows flapping on the ground in a cloud of dust. Did you know they were getting clean? The dust soaks up extra oil and smothers some parasites, too. Water baths also work.

Preening is a full-time job for many birds, including this barn owl.

It might not seem like it, but these house sparrows are getting clean.

Feeding

Eating, hunting, and searching for food (foraging) takes up a lot of a bird's time. And there are almost as many ways to eat as there are types of birds—knowing how birds eat is a behavior clue useful in identifying them.

SWALLOWS eat on the wing, snatching bug after bug right out of the air without stopping.

FLYCATCHERS also eat flying insects but they hunt from a perch, darting out after dinner.

GOLDFINCHES sit and sway on top of thistle heads while pulling out and eating tiny seeds.

BLUE JAYS will pick up an acorn and take it to a high branch before cracking open its snack.

RAPTORS such as hawks, falcons, osprey, and others scan for prey, then ambush it with a fast dive, seizing it with strong feet.

WATERBIRDS such as ducks, herons, terns, and others find food in all sorts of ways. Some ducks go tail up in the water to get at plants near the bottom. Herons stalk and grab fish and frogs with their long beaks, while shorebirds run and probe sand for worms and little crabs.

Flying and Moving

Some birds fly in ways that give themselves away. Goldfinches and woodpeckers fly in an up-and-down wave pattern, while crows fly straight and fast. ("As the crow flies" is the quickest route!) You can often identify a large faraway bird high overhead by how it flies. Great blue herons fly high above with impossibly slow flaps. Turkey vultures soar with their wings in a teetering V shape, while hawks glide steady with wings outstretched.

Crows (top) flap steady and fly straight, but ravens (bottom) flap and glide and/or flap and soar.

Finches (top) bounce up as they flap, glide down like riding a wave, and then flap to bounce back up. Woodpeckers (bottom) also flap and glide in a wave-like pattern.

Most birds spend some time on the ground or in trees. Watching how a bird gets around when it's not flying can help identify it, too. Pigeons walk one foot at a time, while wrens hop from place to place with tails up. Woodpeckers scale up tree trunks with ease, but nut-hatches can scurry down clinging to bark headfirst, too.

Courtship

PEACOCK

Courtship behavior is some of the most fascinating bird behavior there is. Male turkeys and peacocks strut with tails unfurled to show off for mates. Cranes start the breeding season with a court-ship dance of leaps and flaps all while loudly bugling. Bald eagle pairs do a dangerous-looking flight that takes them tumbling through the air with talons locked in freefall. Complicated repeated series of moves involving bobbing heads, smacking bills, bumping chests, wrap-ping necks, flailing legs, daredevil dives, and flapping wings are all ways to show you're in the know and will be a good mate.

SANDHILL CRANES

Communication

Every species of bird has its own distinct songs and sounds. Listening and learning different bird songs will help you find birds, identify them, and know what they're up to. Plus you can enjoy bird songs without *seeing* a single bird. Birds make a huge variety of sounds—whistling, twittering, crackling, cawing, screaming, gurgling, etc. And individual birds make more than one kind of sound.

A bird's song is usually a series of notes. Male birds sing their songs to attract females and warn away other males from their territory. That lovely sound can mean, "Get out! NOW!" A bird's call is usually less musical and shorter. A bird calls to let its mate, young, or other flock members know where it is.

BLACKBURNIAN WARBLER

An alarm call warns of danger, like a hawk, cat, snake, or other predator. Singing isn't the only way birds communicate. Woodpeckers warn and woo by using their beak like a drumstick on hollow trees. It's loud! Male ruffed grouse run off rivals with noisy beating wings.

Learning bird songs takes practice. There's no one right way to do it. Watching a bird and listening to it sing at the same time connects one to the other for many birders.

Woodpeckers, like this pileated woodpecker, announce their territory by drumming on hollow trees.

There are many memory-helping, or mnemonic, phrases that birders use to describe and remember bird songs. Have you heard of any of these?

AMERICAN ROBIN

cheerily, cheer up, cheer up, cheerily, cheer up

CAROLINA WREN

tea-kettle, tea-kettle, tea-kettle

BARRED OWL

who-cooks-for-you, who-cooks-for-you-all

CHICKADEE

chickadee-dee-dee

BLUE JAY

jay, jay

EASTERN BLUEBIRD

cheer, cheerful charmer

INDIGO BUNTING
fire, fire, where, where, here, here

TUFTED TITMOUSE
peter, peter, peter

NORTHERN CARDINAL
what-cheer, cheer, cheer

WOOD THRUSH
eee-oh-lay

OVENBIRD
teacher, teacher, teacher

YELLOWTHROAT
witchety, witchety, witchety

RED-WINGED BLACKBIRD
con-clare-eee

YELLOW WARBLER
sweet, sweet, sweet, little-more-sweet

Everyone learns differently and many of us benefit from more than one way, so try it all—song descriptions, mnemonic phrases, recordings, and live sightings. Luckily, looking for birds singing is a great way to spend a nice day.

Bird Behavior Evidence

Some bird behaviors don't have to be seen directly to know they happened. Here are some search-worthy examples of what active birds have left behind. Look for:

CHIPS OF WOOD under a dead tree where a woodpecker's been working.

SHELLED SEEDS AND CRACKED NUTS near a favorite eating perch.

BIRD TRACKS in snow and mud. Look for webbed feet near water.

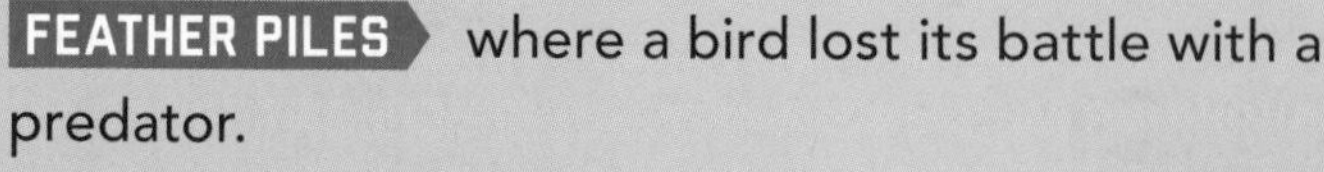

FEATHER PILES where a bird lost its battle with a predator.

WHITE LIQUID POOP on branches or on the ground under a roost. Watch out!

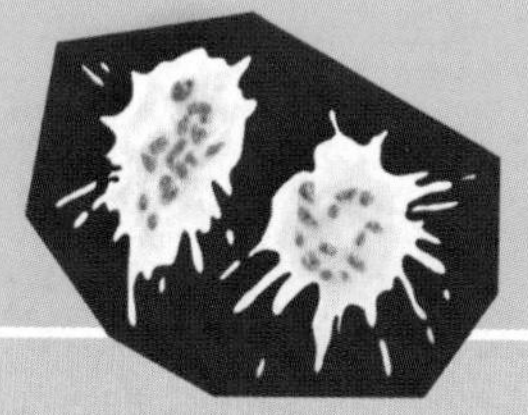

Nests of Many Sorts

A bird nest can be as elaborate as a hanging globe of expertly woven blades of grass, or a simple, scraped-out hole in the dirt. Suitable

places to nest, as well as the materials—such as sticks, leaves, and mud—to build them are part of a bird's habitat needs. Hummingbird nests are the size of doll's teacup and are made of lichen stuck together with collected spiderwebs. Hawk nests are stacks of sticks high up on trees or towers. Barn swallows create cups of dried mud under eaves. Wrens' nests are moss and grass famously shoved in all sorts of funny places, including hats and boots left on porches.

WREN

Nests tell you a lot about how and where a bird lives. Here are a few general kinds of nests and the birds who make them:

BURROWS Holes dug sideways into a stream or river's bank are safe spots to nest. (Racoons and jays can't sneak in.) Bank swallows and kingfishers build these. Burrowing owls take over prairie dog holes for nests.

CAVITIES Holes, or cavities, in trees are nesting spots protected from the weather and predators for chickadees, bluebirds, owls, and some warblers as well as the woodpeckers who pecked them out.

DOWNY
WOODPECKER

PLATFORMS Open, flat piles of sticks and grasses on the ground, in marshes, or high in a tree make wide nests that are easy for large birds to land on.

CUPS Nests with a cup shape of one sort or another suit many birds. Swifts make them of mud stuck to rock (or inside chimneys!). Robins make cups of easy-to-find twigs, mud, and grasses.

SCRAPES Simple shallow depressions in leaf litter, sand, or gravel keep the eggs from rolling away. Gulls and shorebirds including killdeer, vultures, and nighthawks nest like this.

KILLDEER

Nest Spotting Tips

After trees have lost their leaves in the fall, old bird nests are easy to spot. But many birds try to hide their nests during nesting season (spring and summer), so spotting them can be tough. Here's some help:

- Make sure you watch from a safe distance using binoculars. Being too close not only disturbs the parents, but your smell can attract predators like raccoons to the nest. Back off and use binocs!
- Listen for squawking, peeping, begging, noisy nestlings. Follow your ears.
- Stalk the parents. See a bird flying by with grass or twigs in its beak? Watch where it's going. Same thing if a bird's carrying caterpillars or other bugs. (Baby food!)
- Monitor tree holes. Cavity nests are prime real estate. Watch them to see who's coming and going.
- Look for eggshells on the ground. New hatchlings could be nearby, though some bird parents carry them away to fool predators (and nest watchers).

BALTIMORE ORIOLE

TRY IT → Spot Birds Behaving

Bird behavior is likely happening all around you! Go see some.

AMERICAN ROBIN

WHAT YOU'LL NEED

- a pencil or pen, binocs (optional)

STEP 1 Go for a walk or hike outdoors and have a look for yourself.

STEP 2 Check off those bird behaviors you observe, and note any details about what you saw (or heard!).

- ☐ preening ....
- ☐ flocking ....
- ☐ alarm calling ....
- ☐ flying in an up-and-down wave pattern ....
....
- ☐ bathing ....

- [ ] soaring high ....

- [ ] courting ....

- [ ] singing ....

- [ ] walking ....

- [ ] climbing ....

- [ ] flapping wings on ground ....

- [ ] calling ....

STEP 3 Can you identify any of the birds whose behaviors you witnessed? Do you know their names? Look for them in the Bird Identification section (pages 86–187). Make a positive ID? Way to go! Check off its I SAW IT! box and fill in the blanks.

TRACK IT ↘ Find Flight and Flock Patterns

STEP 1 Observe how birds are flying. Can you spot any that fly with a particular pattern? Such as the forward-and-back zippy flight of a hummingbird or the V-shaped soar of a vulture? How about bird flock patterns, like the fast-paced jumble of chimney swifts or slow organized V of geese?

STEP 2 When you see a flight or flock pattern, track it on the page below. Use the blank space to draw the pattern.

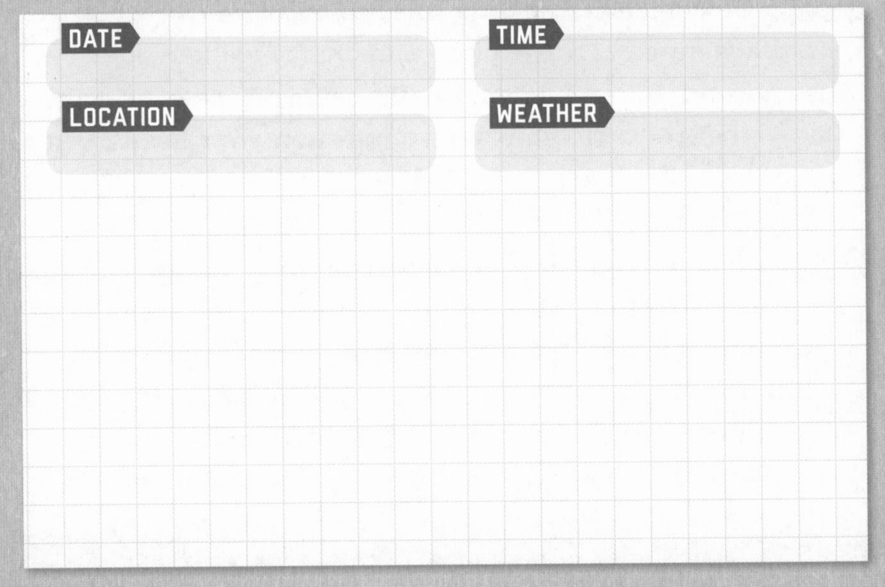

STEP 3 Flight and flock patterns can help with ID. Did you recognize the bird you tracked? Look for them in the Bird Identification section (pages 86–187). Check off its I SAW IT! box and fill in the blanks. Use the space for notes to include the flight pattern you saw.

TRY IT → Name That Tune

Ready to learn bird songs? You already know some. A duck says quack and a pigeon sings *coo-coo*. You likely recognize other bird songs, too, but just haven't connected them to a bird name yet. So let's get started!

WOOD DUCKS

WHAT YOU'LL NEED

- binocs, a bird field guide, and a pencil or pen

STEP 1 Go someplace outside where birds are that's quiet. (Traffic noise and chatty walkers are distracting.) Sit and listen. Really listen. Close your eyes. Cup your ears with your hands and turn your head right to left, to zero in on birds singing.

STEP 2 Pick out one particular call or song that is repeating. Actively listen to it. Focus on its pattern of beats, string of notes, and whether the tune changes or gets louder. Can you hum, sing, or whistle it? Describe the sound to yourself as you listen. How would you explain to someone what it sounds like?

STEP 3 Pick up the binoculars and try to spot the bird that's singing. If you're lucky enough to get a good look, observe it singing. Focus again on the sound, its pattern, notes, and tone, and describe it to yourself again.

STEP 4 Answer these questions about the song:

→ What is its rhythm? Tap it out.

→ What is the melody? Hum or sing it to yourself.

→ Does it get louder or softer?

→ Does it change in tone? How?

STEP 5 Write down your description of the song. There's no right or wrong, just what it sounds like to you. Some examples of ways to describe bird songs:

- Two fast chirps and a long buzz or *CHIP, CHIP, BURRRRRRRR*
- Flute sound that gets louder or *twee-dle, twee-DLE, TWEE-DLE*.
- One high, one low, then long note or *chip-churr-CHEEEE*.

STEP 6 Did you recognize the bird you tracked? Look for it in the Bird Identification section (pages 86–187). Make a positive ID? Nice work! Check off its I SAW IT! box and fill in the blanks. You can use the space for notes to describe its song.

TRACK IT ↘ What's It Saying?

WHAT YOU'LL NEED

➢ a pencil or pen, paper

HOUSE FINCH

Many of the birdsong mnemonic phrases are pretty old-fashioned. One for the white-throated sparrow is *Old Sam Peabody, Peabody, Peabody.* (Who the heck is Sam Peabody?) Reinforce your own memory by inventing phrases for your favorite bird songs or songs you hear around you a lot. For example, here's one for a white-throated sparrow that's just traveling through on its migration north: *I'm-heeeerrrreeee, but-not-for-long, not-for-long, not-for-long!*

TAKE IT TO THE NEXT LEVEL ↗

Record Some Songsters

Do you have a voice recording device or app? Recording bird songs can help you remember them. It's also helpful to have a recording if you want to identify the bird later on—you can match it to a bird song from an online library or ask an expert.

INDIGO BUNTING

TRY IT → Scout Nest Sites

Start nest watching by first noticing where nests are.

BLUE-GRAY GNATCATCHER

WHAT YOU'LL NEED

- a pencil or pen, binocs (optional)

Check off the places where you've seen bird nests:

- ☐ tree branches
- ☐ tree trunk holes
- ☐ under bridges
- ☐ on the ground
- ☐ in bushes
- ☐ under roof eaves
- ☐ river or stream banks
- ☐ tops of utility poles

Where else? ....

GRAY CATBIRD

I DID IT! DATE:

TRACK IT ↘ Spot an Abandoned Nest

Nests are amazing pieces of architecture. Take a closer look!

WHAT YOU'LL NEED

➢ a pencil or pen, an abandoned nest

STEP 1 Find an *abandoned* nest. Look in fall or winter when birds are no longer breeding, or find one that obviously fell out of a tree. (If in doubt, let it be!)

STEP 2 Describe the nest and draw it below.

What shape is it?

What is it made of?

How do you think the bird made it?

DATE

TIME

LOCATION

WEATHER

Draw the nest.

TRACK IT ↘ Active Bird Nest Watch

Find an active nest you can easily observe with zero disturbance (none!) of the birds, and dig deeper into why it works as a place to raise young.

WHAT YOU'LL NEED

- a pencil or pen, binocs (optional)

STEP 1 Be as exact as you can with the Nest Site description. If it's in a tree or under a bridge, how far off the ground is it? Is it wedged between branches or sitting atop a pole?

STEP 2 Check off *all* the materials you see in the nest. A single hair counts!

STEP 3 Draw the nest in the blank space provided on the next page. Pay attention to the nest's shape, depth, and width.

STEP 4 Are there eggs? What can you compare their size to: peas, grapes, cherry tomatoes? Color includes any patterns like speckles or spots.

STEP 5 Can you identify the parents? Remember that male and female birds can look different. Look for them in the Bird Identification section (pages 86–187). Make a positive ID? Fabulous work! Check off its I SAW IT! box and fill in the blanks.

NORTHERN CARDINALS

DATE

TIME

LOCATION

WEATHER

Nest Site:

Nest Shape:

- [ ] cup
- [ ] dome
- [ ] platform
- [ ] other

Nest Material:

- [ ] dried mud
- [ ] twigs or sticks
- [ ] grass
- [ ] leaves or pine needles
- [ ] moss
- [ ] vines
- [ ] hair or fur
- [ ] feathers
- [ ] plant fibers like milkweed or thistle fluff
- [ ] trash (string, plastic, fabric)

Anything else? ....

Draw the nest.

Eggs? ☐ yes ☐ no How many?

Size:

Color:

Describe the parents.

STEP 6 If the nest has eggs and parents are around, you're in luck! Try to visit every day, creating a daily journal entry about the state of the eggs, like the one below. Under Egg Status, write down when any eggs are showing cracks. Drawing them helps you keep track of their progress. Once they hatch, keep journaling! Observations is a place to note whether the parents are around, when the chicks get feathers, when they fledge, etc.

DATE

TIME

Egg/chick count:

Egg/chick status:

Observations:

I DID IT! DATE:

CHAPTER 7

How Birds Live

Want to see a particular bird? Try the kitchen. Or at least the bird version of it. Birds eat a lot, so looking for a bird where its food is plentiful is always a good bird-watching strategy. Birds are hungry because, like you, they are warm-blooded animals. Their bodies maintain a steady temperature by continually burning fuel—the food they eat. Flying is also a high-energy activity. All this adds up to a lot of chow time. Eating like a bird actually means eating a lot! A little chickadee eats about a third of its weight each day to power its on-the-wing lifestyle. For a hundred-pound human that'd be about a dozen extra-large pizzas. Birds need a lot of food to survive.

Beaks and Feet

The mouths of birds—beaks or bills—are toothless. Birds don't chew, they gulp. The food gets ground up by a gizzard, a muscular part of the stomach (some birds swallow gravel to help their gizzards grind up food). Bird beaks come in a fascinating variety of shapes, sizes, and colors (think toucan!) that can help you identify both their group and the species. Beaks are more than mouths. Many birds use beaks like a hand or arm to help gather sticks and nesting material, dig holes, groom, and even climb trees.

YELLOW-RUMPED WARBLER

Feet also come in handy for armless birds. And feet are another bird body part perfectly adapted to the lifestyle of their owner and a great clue to identification. Penguin feet are good for swimming, ostrich feet are made to run, and an eagle has strong grabbing feet armed with sharp talons.

Have you ever wondered why a bird asleep on a branch doesn't fall off? When a bird bends its knees and ankles its feet automatically lock into a strong grasp. Only straightening the legs unlocks the feet's hold on their perch.

Bird Sense

Birds in general have large eyes, excellent eyesight, and see in full color. Vision is the sense most birds depend on to find food, fly safely, and seek out mates. But birds also hear well, and listening is a big part of bird communication. Their ears are on the sides of the head under their feathers. While birds have taste buds and noses, in general they aren't good smellers. Only a few kinds of the birds—ocean albatrosses and carrion-eating vultures—have an excellent sense of smell. Poor smelling and tasting ability are why adding pepper powder to birdseed keeps squirrels away from feeders, but not birds. It's also why putting a baby bird back into its nest is safe. The parents won't smell your scent.

TRY IT → Go on Beak Watch

A bird's beak or bill (same thing) is a tool for eating. What that bird-feeding tool is shaped like and how it works depends on the food it's after.

WHAT YOU'LL NEED

➢ a pencil or pen, binocs

TERN LOON HERON KINGFISHER

STEP 1 Go bird-watching! A place with a variety of habitats (water, woods, and open country) is ideal.

STEP 2 Check off these bird beaks when you see them in action:

- ☐ wide, flat, scooping and straining beak
- ☐ curved, thick, flesh-ripping beak
- ☐ thin, straw-like, nectar-sipping beak
- ☐ triangular, chunky, seed-crushing beak
- ☐ narrow, pointy, bug-catching beak
- ☐ sharp, strong, pointed, wood-chipping beak

STEP 3 Could you identify any of the birds whose beaks you saw at work? Look for them in the Bird Identification section (pages 86–187). Make a positive ID? Congrats! Check off its I SAW IT! box and fill in the blanks.

TRY IT → Go on Foot Watch

Bird feet are built for how a bird gets around on land, in water, or both. They can also be tools for hunting and roosting.

WHAT YOU'LL NEED

- a pencil or pen

STEP 1 Go bird-watching! A place with a variety of habitats (water, woods, and open country) is ideal.

STEP 2 Check off these bird feet when you see them doing what they do best:

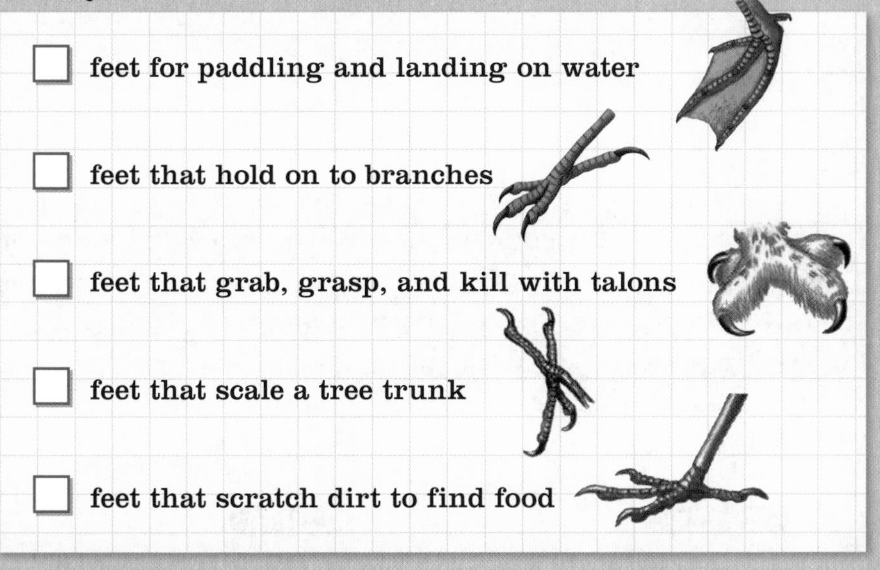

- ☐ feet for paddling and landing on water
- ☐ feet that hold on to branches
- ☐ feet that grab, grasp, and kill with talons
- ☐ feet that scale a tree trunk
- ☐ feet that scratch dirt to find food

STEP 3 Could you identify any of the birds whose feet you saw at work? Look for them in the Bird Identification section (pages 86–187). Make a positive ID? Congrats to you! Check off its I SAW IT! box and fill in the blanks.

TRACK IT ↘ Sketch a Bird Beak or Foot

Look closely at one of the birds whose beak or feet you checked off.

WHAT YOU'LL NEED

➢ a pencil or pen

STEP 1 Use the blank space below to draw its beak or foot—or both!

STEP 2 Did you recognize the bird whose foot or beak you drew? Look for it in the Bird Identification section (pages 86–187). Make a positive ID? Fabulous work! Check off its I SAW IT! box and fill in the blanks.

DATE

TIME

LOCATION

WEATHER

Did you identify the bird?........................ If so, did the feet and/or beak help in the ID? ....

TAKE IT TO THE NEXT LEVEL ↗

Feeder Watch

Is there a bird feeder at your house, or at a nearby park or nature center? If not, you can make your own by putting a little pile of mixed wild birdseed out on a stump or picnic table. Then watch the birds come and chow down. Look for:

- → **Which kinds of birds eat the smaller seeds?**
- → **Who only goes for the sunflower seeds?**
- → **Do the seed sizes match the beak sizes, or not really?**
- → **Which birds like to "dine in," staying right by the food while eating?**
- → **Who likes "takeout," grabbing a seed and then flying off to a perch to eat?**

TAKE IT TO THE NEXT LEVEL ↗

Build a Bird Profile

Find out how your bird knowledge stacks up by putting it to use on one single kind of bird. Choose a species of bird you can observe in nature. Then build a complete profile of the bird.

What is the name of the bird?

Draw its silhouette.

What's its size?

What's its overall color pattern?

Does it have field marks on its head or wing?

What are its feet and beak like? Draw them.

What does it eat?

How does it fly?

What's its song like?

Its call?

What kinds of behavior have you observed the bird doing?

Describe its habitat.

Where, exactly, have you seen it?

You're on the road to ornithology now!

BIRD IDENTIFICATION

Welcome to your guide to bird identification.
Here are some tips to get started:

COMMON NEIGHBORHOOD BIRDS

Birds are mostly grouped by habitat, but there is a section of common neighborhood birds to start.

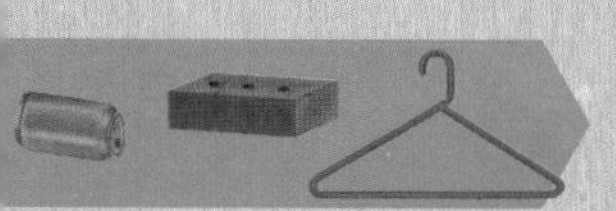

The silhouette summarizes both SHAPE and relative size compared to a sparrow, a robin, or a crow. Each bird's silhouette is placed between two of these sizing birds so you can estimate its size.

A sparrow is about 6 inches (15 cm) beak to tail tip, or about an inch longer than a soda can on its side. A robin is about 8½ inches (21½ cm), or a little bit longer than a brick. And a crow is about 17 inches (43 cm), or about as long as a clothes hanger is wide.

Details about songs, singing, and calls are often under BEHAVIOR.

Read the key when using range maps to see where birds are during different times of the year.

See page 51 for head and wing field marks.

Blue Jay

(Cyanocitta cristata)

SIZE 11 inches (28 cm) beak to tail

SHAPE Large perching bird with crest on head and wide tail with rounded end.

COLOR White underneath and blue, black, and white pattern above.

BEHAVIOR Makes variety of loud, noisy calls while perched.

HABITAT Forest edges, parks, backyards.

POINT OF FACT A crest up means alert and ready for action! Jays lower their crests when tending to their young, hanging with family, and relaxing.

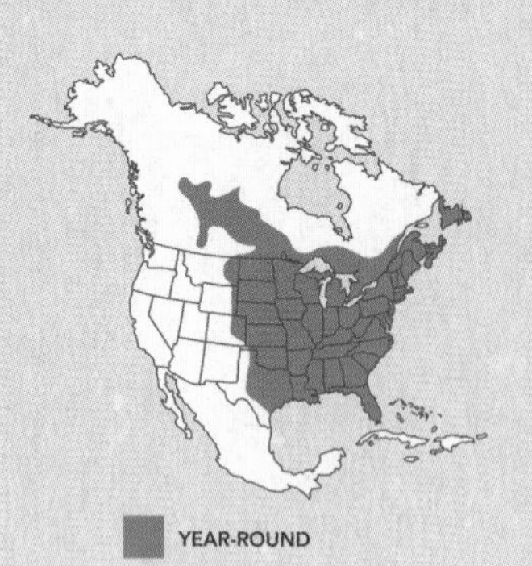

I SAW IT!

WHEN I SAW IT
DATE

WHERE I SAW IT
SPECIFIC LOCATION, INCLUDE STATE

WHAT IT WAS DOING

NOTES

American Goldfinch

(Spinus tristis)

Black forehead cap.

SIZE 4¼ inches (11½ cm) beak to tail

MALE

White markings on black wings.

SHAPE Small perching bird with a small head, a thick, stubby beak, long wings, and a notched tail.

COLOR Bright yellow with a black cap and wings. Females and wintertime males are a dull yellow to olive with no cap.

FEMALE

BEHAVIOR Small flocks cling and balance like acrobats on seedy plants.

HABITAT Backyards, roadsides, weedy fields.

POINT OF FACT Flies in a bouncy, up-and-down wave pattern while calling *po-ta-to-chip* to other goldfinches.

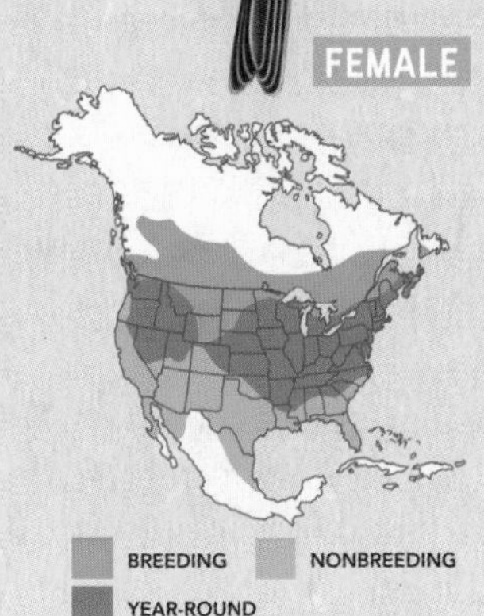

I SAW IT!

WHEN I SAW IT

DATE

WHERE I SAW IT

SPECIFIC LOCATION, INCLUDE STATE

WHAT IT WAS DOING

NOTES

American Robin

(Turdus migratorius)

Broken, white eye ring.

Orange chest.

SIZE 8½ inches (21½ cm) beak to tail

SHAPE Large perching bird with a round body, long legs, and a longish tail.

COLOR Gray-brown back and wings, a dark brown head, and an orange chest. White lower belly patch. The color of females is sometimes paler overall.

BEHAVIOR Confidently hops around on the ground to search for food.

HABITAT Gardens, parks, yards, fields, woodlands, and forest edges.

POINT OF FACT An unreliable first sign of spring, since not all American robins migrate elsewhere in the winter. State bird of Connecticut, Michigan, and Wisconsin.

I SAW IT!

WHEN I SAW IT

DATE

WHERE I SAW IT

SPECIFIC LOCATION, INCLUDE STATE

WHAT IT WAS DOING

NOTES

COMMON NEIGHBORHOOD BIRDS

European Starling

(Sturnus vulgaris)

SIZE 6 inches (15 cm) beak to tail

SHAPE Husky perching bird with a short tail and short, pointy wings.

COLOR Glossy black with a purple-green iridescence and a yellow beak in the summer. Spots cover its back and chest in the winter.

BEHAVIOR Flocks in big, noisy groups, sits on wires, and makes whistles and other sounds to other birds.

HABITAT Cities, fields, lawns.

POINT OF FACT When flying, its wings look like small, four-point stars. That's where the name "starling" comes from.

I SAW IT!

WHEN I SAW IT

DATE

WHERE I SAW IT

SPECIFIC LOCATION, INCLUDE STATE

WHAT IT WAS DOING

NOTES

COMMON NEIGHBORHOOD BIRDS

Gray Catbird

(Dumetella carolinensis)

SIZE 7¼ inches (18½ cm) beak to tail

SHAPE Medium-size, slender perching bird with a long, rounded tail and a narrow beak.

COLOR Slate gray.

BEHAVIOR Fast branch-to-branch hopping with tail flicks and quick, low flights over bushes.

HABITAT Forest edges, overgrown fields, bushy streamside areas.

POINT OF FACT It sings like a meowing cat for up to ten minutes. It also sings made-up songs that it samples from other birds.

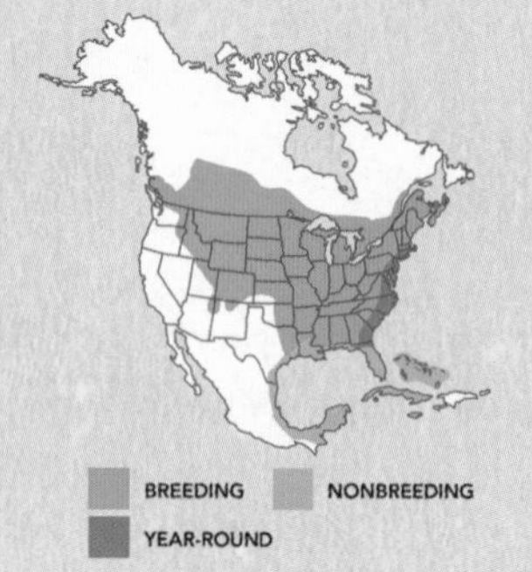

I SAW IT!

WHEN I SAW IT

DATE

WHERE I SAW IT

SPECIFIC LOCATION, INCLUDE STATE

WHAT IT WAS DOING

NOTES

House Sparrow

(Passer domesticus)

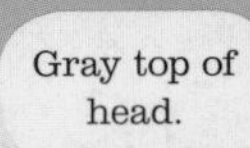

SIZE 6 inches (15 cm) beak to tail

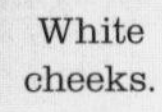

Black bib.

MALE

FEMALE

SHAPE Beefy, small perching bird with a large head, a short tail, and a stubby beak.

COLOR Brown, black, and white streaky wings and back with a red-brown and gray head. Females are a plain tan overall with gray coloring underneath.

BEHAVIOR Noisy flocks search for food, and nest in homes and buildings.

YEAR-ROUND

HABITAT City streets, busy parks, roofs, backyards.

POINT OF FACT Eats grains, seeds, insects, as well as pizza crust, hot dog buns, and other tossed-out human leftovers in the summer to feed to its young.

WHEN I SAW IT

DATE

WHERE I SAW IT

SPECIFIC LOCATION, INCLUDE STATE

WHAT IT WAS DOING

NOTES

Mallard

(Anas platyrhynchos)

FEMALE

White neck band.

MALE

White-bordered blue wing patch.

SIZE 23 inches (58½ cm) beak to tail

SHAPE Large duck with a long body, a round head, and a flat, wide beak.

COLOR A rusty brown chest, a gray body, a dark, iridescent green head, and a yellow beak. Females are a splotchy brown with dark orange beaks.

BEHAVIOR Tips tail up in the water while feeding.

HABITAT Lakes, ponds, and wetlands, including in city parks.

POINT OF FACT Mallards aren't divers; they dabble for seeds and aquatic plants by tipping forward into the water.

I SAW IT!

WHEN I SAW IT
DATE

WHERE I SAW IT
SPECIFIC LOCATION, INCLUDE STATE

WHAT IT WAS DOING

NOTES

Mourning Dove

(Zenaida macroura)

SIZE 11 inches (28 cm) beak to tail

SHAPE Slender dove bird with a very small head, short legs, a small beak, and a long tail.

COLOR Buff to light brown bodies and heads. Juveniles have white wing tips and flecks on their faces.

BEHAVIOR Pecks on the ground for seeds and roosts in small flocks on overhead wires.

HABITAT Backyards, fields, deserts, scrublands.

POINT OF FACT Its name comes from its slowly repeating, sad-sounding call: *coo-EEE, coo . . . coo.*

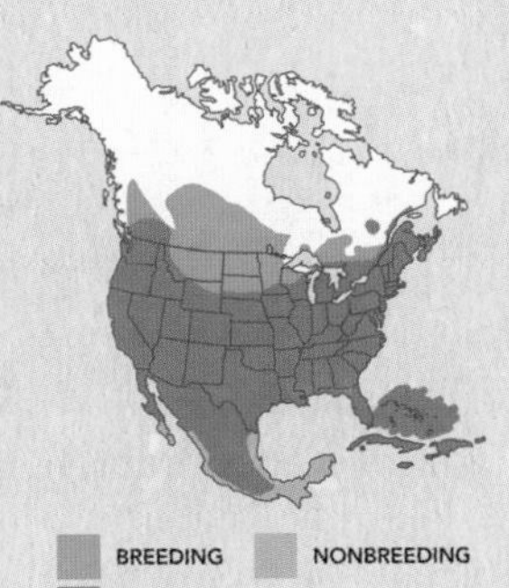

I SAW IT!

WHEN I SAW IT

DATE

WHERE I SAW IT

SPECIFIC LOCATION, INCLUDE STATE

WHAT IT WAS DOING

NOTES

Northern Cardinal

(Cardinalis cardinalis)

Red wings.

MALE

SIZE 8½ inches (21½ cm) beak to tail

Black face around red beak.

SHAPE Medium-large perching bird with a crest on its head, a long tail, and a thick, short beak.

COLOR Males are red. Females are a pale yellow-brown with a red tint on their crests and wings.

BEHAVIOR Sits in shrubs and on low tree branches and forages on the ground.

FEMALE

HABITAT Backyards, parks, forest edges, woodlots.

POINT OF FACT Cardinals are a popular state bird! Illinois, Indiana, Kentucky, North Carolina, Ohio, Virginia, and West Virginia all chose them as their official bird mascots.

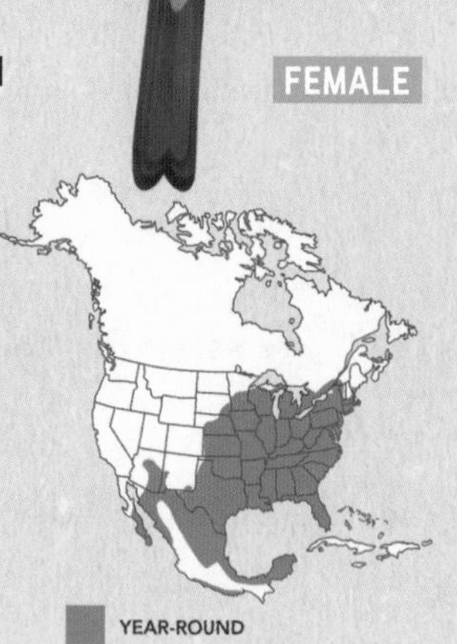

I SAW IT!

WHEN I SAW IT

DATE

WHERE I SAW IT

SPECIFIC LOCATION, INCLUDE STATE

WHAT IT WAS DOING

NOTES

COMMON NEIGHBORHOOD BIRDS

Northern Flicker

(Colaptes auratus)

SIZE 11½ inches (29 cm) beak to tail

SHAPE Large woodpecker with a slim, round head, a long, strong beak, and a long, flared tail.

COLOR Light brown overall with black spots and bars on its body, bright red or yellow under its wings and tail, and a white rump patch.

BEHAVIOR Flies, flaps, and glides in an up-and-down pattern. Unlike other woodpeckers, the Flicker spends time on the ground and perched on branches.

HABITAT Woodlands, forest edges, yards, parks, and western forests.

POINT OF FACT State bird of Alabama, where it's often called a yellowhammer woodpecker. Civil War soldiers from the state often wore its feather into battle.

I SAW IT!

WHEN I SAW IT

DATE

WHERE I SAW IT

SPECIFIC LOCATION, INCLUDE STATE

WHAT IT WAS DOING

NOTES

Rock Pigeon

(Columba livia)

Orange eyes.

Black wing bands.

SIZE 13 inches (33 cm) beak to tail

SHAPE Large, plump dove with a small head, short legs, and a wide, rounded tail.

COLOR Varies, but most are a blue-gray with darker heads, iridescent throats, and dark-tipped tails.

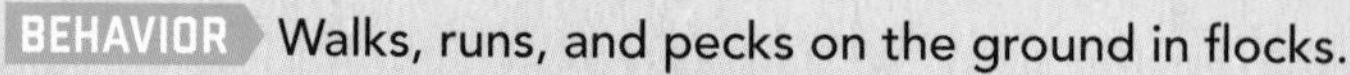

BEHAVIOR Walks, runs, and pecks on the ground in flocks.

HABITAT Cities, towns, parks, farmland, fields, rocky cliffs.

POINT OF FACT Pigeon eyes can sense earth's magnetic field, which helps them find their way home from distant locations.

I SAW IT!

WHEN I SAW IT

DATE

WHERE I SAW IT

SPECIFIC LOCATION, INCLUDE STATE

WHAT IT WAS DOING

NOTES

Northern Mockingbird

(Mimus polyglottos)

SIZE 10 inches (25 cm) beak to tail

SHAPE Medium-size, slender perching bird with a small head, a thin beak, a long tail, and long legs.

COLOR Gray-brown with a pale chest and belly.

BEHAVIOR Sings loudly while sitting out in the open on high fences, poles, eaves, and tree branches.

HABITAT Lawns, backyards, towns, parks, forest edges.

POINT OF FACT It can imitate more than a hundred other bird songs and different sounds. It sings into the night during spring and summer, especially during full moons.

I SAW IT!

WHEN I SAW IT

DATE

WHERE I SAW IT

SPECIFIC LOCATION, INCLUDE STATE

WHAT IT WAS DOING

NOTES

....

....

Ring-billed Gull

(Larus delawarensis)

SIZE 18 inches (45½ cm) beak to tail

SHAPE Medium-size gull with a short beak.

COLOR White head, body, and tail with gray wings and back, yellow legs, and a yellow beak. Juveniles are a splotchy brown and gray with pink legs and beaks.

BEHAVIOR Makes high-pitched squealing cries as it circles and hovers overhead searching for food.

HABITAT Coastal beaches, large lakes, garbage dumps, parking lots.

POINT OF FACT Some people call it the "McGull" because it likes fast-food dumpsters and leftovers.

I SAW IT!

WHEN I SAW IT

DATE

WHERE I SAW IT

SPECIFIC LOCATION, INCLUDE STATE

WHAT IT WAS DOING

NOTES

Red-winged Blackbird

(Agelaius phoeniceus)

SIZE 8 inches (20½ cm) beak to tail

SHAPE Stocky, perching blackbird with a medium-size tail and a slender beak.

COLOR Glossy black with red-and-yellow shoulder patches, a black beak, and black legs. Females are brown and streaky with lighter-colored chests.

BEHAVIOR Sits on high, open perches while screaming out *conk-la-ree*.

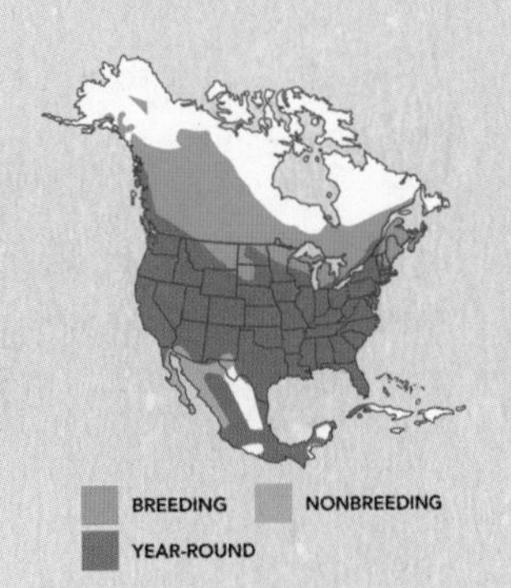

HABITAT Marshes, fields, wet roadsides, and meadows.

POINT OF FACT Males show up early in the spring to set up their territories, singing for hours a day and running off other males to defend their homes.

I SAW IT!

WHEN I SAW IT

DATE

WHERE I SAW IT

SPECIFIC LOCATION, INCLUDE STATE

WHAT IT WAS DOING

NOTES

COMMON NEIGHBORHOOD BIRDS

Red-tailed Hawk

(Buteo jamaicensis)

SIZE 20 inches (51 cm) beak to tail

SHAPE Large hawk with broad, rounded wings and a wide, fanned tail.

COLOR Medium-brown back and wings, with a lighter chest, wing undersides, and tail. Some have a red coloring and others are a darker brown.

BEHAVIOR Soars in wide circles over fields and roosts on high fences and poles.

HABITAT Open forest edges, deserts, grasslands, fields.

POINT OF FACT Scans for mice, squirrels, and other small mammals while soaring overhead, then drops down and grabs its prey with its talons.

I SAW IT!

WHEN I SAW IT
DATE

WHERE I SAW IT
SPECIFIC LOCATION, INCLUDE STATE

WHAT IT WAS DOING

NOTES

Ruby-throated Hummingbird

(Archilochus colubris)

SIZE 3 inches (7½ cm) beak to tail

sparrow robin

SHAPE Small hummingbird with a long, thin beak and short wings.

COLOR Shiny green on its back and head with white-gray underparts. Males have bright red iridescent throats. Females and juveniles have white throats.

BEHAVIOR Zips up, down, backward, and forward with humming wings.

HABITAT Parks, gardens, backyards, forest edges, open woodlands.

POINT OF FACT The humming sound it makes comes from beating its wings more than 50 times a second.

I SAW IT!

WHEN I SAW IT
DATE

WHERE I SAW IT
SPECIFIC LOCATION, INCLUDE STATE

WHAT IT WAS DOING

NOTES

Black-capped Chickadee

(Poecile atricapillus)

SIZE 5¼ inches (13½ cm) beak to tail

Black cap and bib.

White cheeks.

SHAPE Small, round perching bird with a large head, a short neck, a long, narrow tail, and a short beak.

COLOR Gray back and wings, cream undersides, black head.

BEHAVIOR Bouncy, flocking flier that is curious and chatty.

HABITAT Forests and woodlands, neighborhoods, parks.

POINT OF FACT Their calls are complex and help their flock communicate, stay together, and warn of approaching predators.

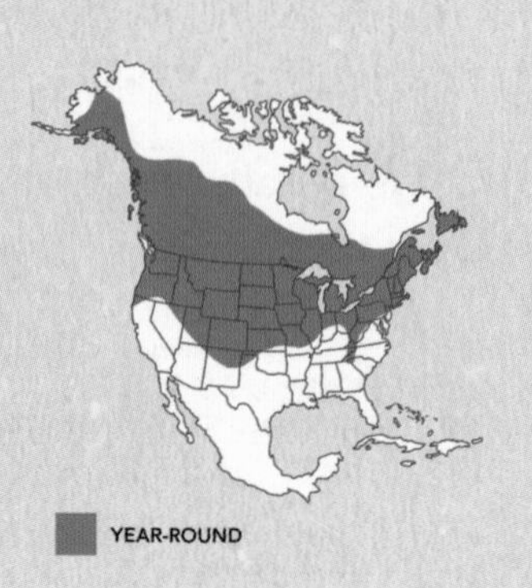

YEAR-ROUND

I SAW IT!

WHEN I SAW IT

DATE

WHERE I SAW IT

SPECIFIC LOCATION, INCLUDE STATE

WHAT IT WAS DOING

NOTES

Song Sparrow

(Melospiza melodia)

Stripes on crown.

Large spot-like streak on chest.

Rusty eyeline.

SIZE 6 inches (15 cm) beak to tail

SHAPE Medium-size, chunky sparrow with a short beak and a long, rounded tail.

COLOR Streaky brown, gray, and cream.

BEHAVIOR Perches on low branches or shrubs, leaning back and belting out a long song of clattering notes and trills. During flight, it pumps its tail downward and makes short, fluttering movements through bushes and dense branches.

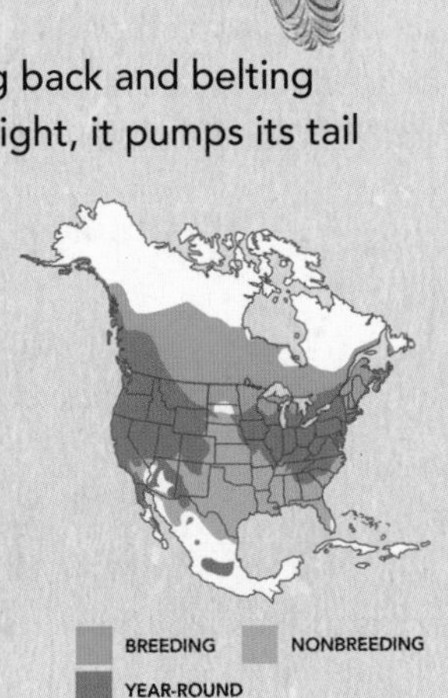

HABITAT Brushy fields, forest edges, backyards, and parks.

POINT OF FACT Females choose males that not only sing well, but are good song learners.

I SAW IT!

WHEN I SAW IT

DATE

WHERE I SAW IT

SPECIFIC LOCATION, INCLUDE STATE

WHAT IT WAS DOING

NOTES

Common Grackle

(Quiscalus quiscula)

SIZE 12 inches (30½ cm) beak to tail

SHAPE Large blackbird with long legs, a long tail, and short wings.

COLOR Appears as black from a distance, but actually has iridescent, bronze tones on its body and deep blue on its head. Females are less glossy.

BEHAVIOR Flies, forages, and roosts in large, noisy flocks.

HABITAT Farm fields, city parks, backyards, forest edges.

POINT OF FACT Eats it all, including corn, acorns, bugs, bats, and sparrows.

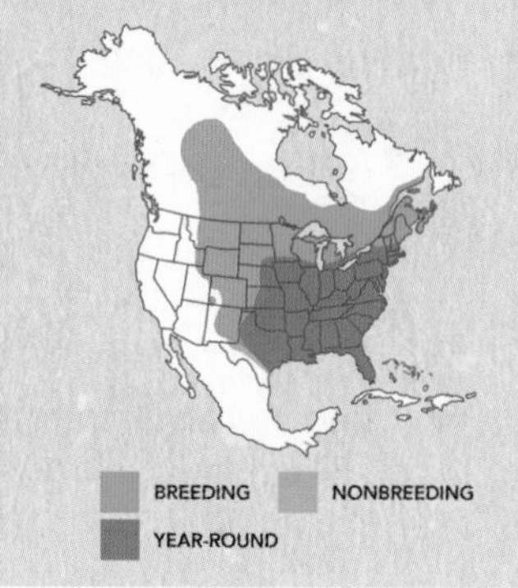

I SAW IT!

WHEN I SAW IT
DATE

WHERE I SAW IT
SPECIFIC LOCATION, INCLUDE STATE

WHAT IT WAS DOING

NOTES

American Crow

(Corvus brachyrhynchos)

SIZE 17 inches (43 cm) beak to tail

SHAPE Large perching bird with a short tail, long legs, a thick neck, and a beefy beak.

COLOR Glossy black all over with black legs and a black beak.

BEHAVIOR Flaps in flight as it yells *caw-caw-CAW*.

HABITAT Open woodlands, fields, towns, roadsides.

POINT OF FACT Intelligent and able to learn, it can count, mimic voices, and make useful tools like whittling down a stick to poke it into a hole full of bugs.

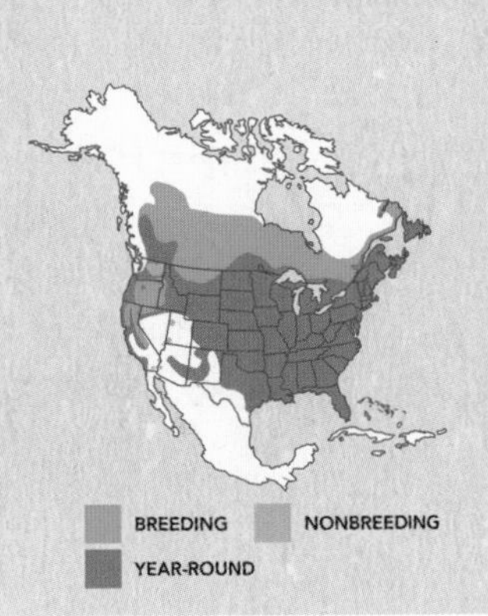

I SAW IT!

WHEN I SAW IT

DATE

WHERE I SAW IT

SPECIFIC LOCATION, INCLUDE STATE

WHAT IT WAS DOING

NOTES

....

....

COMMON NEIGHBORHOOD BIRDS

Chimney Swift

(Chaetura pelagica)

SIZE 5 inches (12½ cm) beak to tail

SHAPE Small perching bird with a tubular body, a short neck, a short tail, and long, curved, narrow wings.

COLOR Dark gray-brown overall with a pale throat area.

BEHAVIOR Flies quickly and acrobatically with fast flaps, short glides, and twisting turns while making a high chattering call.

HABITAT Cities, towns, parks, farmland.

POINT OF FACT Flocks gather near a roost site at dusk, swirling together overhead before dropping into a chimney, cave, or hollow tree for the night.

I SAW IT!

WHEN I SAW IT
DATE

WHERE I SAW IT
SPECIFIC LOCATION, INCLUDE STATE

WHAT IT WAS DOING

NOTES

DESERT BIRDS

Harris's Hawk

(Parabuteo unicinctus)

SIZE 20 inches (51 cm) beak to tail

SHAPE Large hawk with long legs and a long tail.

COLOR Dark and rusty brown overall, a white rump, and yellow legs. Juveniles are pale and splotchy underneath.

BEHAVIOR Soars while fanning its tail and harshly crying *raaaack*.

HABITAT Cactus-covered desert, mesquite scrubland, neighborhoods.

POINT OF FACT It is North America's most social raptor. Pairs or small groups hunt rabbits, ground squirrels, and other rodents as a team. Individuals take turns chasing surrounded prey.

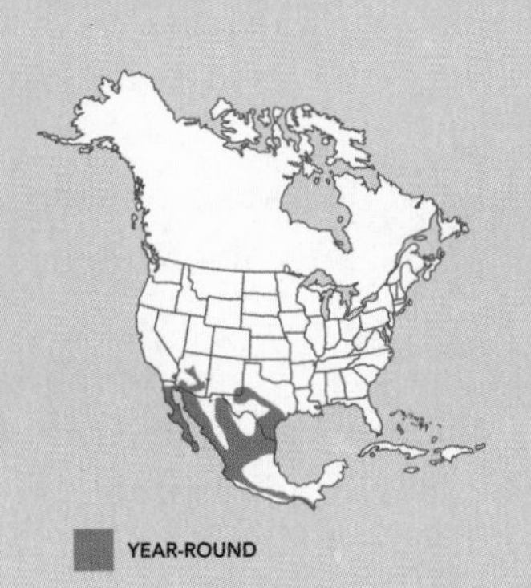

YEAR-ROUND

I SAW IT!

WHEN I SAW IT

DATE

WHERE I SAW IT

SPECIFIC LOCATION, INCLUDE STATE

WHAT IT WAS DOING

NOTES

Cactus Wren

(Campylorhynchus brunneicapillus)

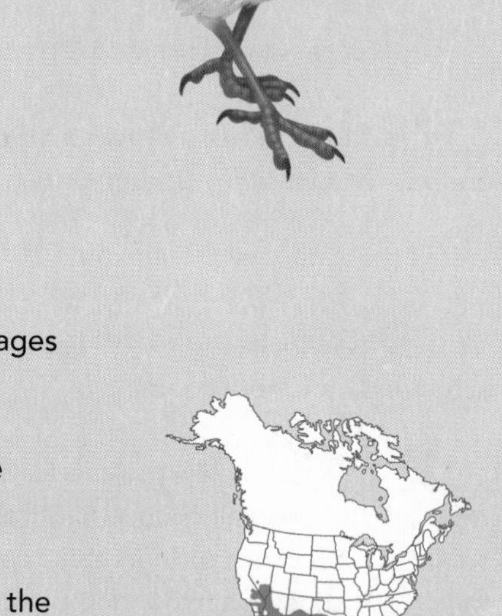

SIZE 7¾ inches (19½ cm) beak to tail

SHAPE Large, chunky wren with a long, rounded tail, and a thick beak.

COLOR Wings, head, and back are brown with light streaks. Its belly and chest are cream with dark speckles and streaks.

BEHAVIOR Perches on cacti and noisily forages out in the open by hopping and scolding.

HABITAT Cactus-covered desert, mesquite scrubland, neighborhoods.

POINT OF FACT A wren gets its water from the spiders, insects, and cactus fruits it eats. It rarely drinks for this reason!

WHEN I SAW IT

DATE

WHERE I SAW IT

SPECIFIC LOCATION, INCLUDE STATE

WHAT IT WAS DOING

NOTES

White-winged Dove

(Zenaida asiatica)

SIZE 11 inches (28 cm) beak to tail

SHAPE Plump dove with a small head, a thin beak, and a square-tipped tail.

COLOR Light brown with white wing patches and white-tipped tail.

BEHAVIOR Forages on the ground and flocks in groups that call a low owl-like *whoo-OOO-oo, ooo-oo.*

HABITAT Desert scrub and cacti lands, backyards, parks.

POINT OF FACT Parent doves feed their nestlings crop milk. It's a nutritious milky mush that builds up on the lining of the crop, the food-stashing part of a bird's throat. Doves vomit up crop milk to feed the babies.

I SAW IT!

WHEN I SAW IT

DATE

WHERE I SAW IT

SPECIFIC LOCATION, INCLUDE STATE

WHAT IT WAS DOING

NOTES

Costa's Hummingbird

(Calypte costae)

MALE

Purple throat patch that flares out.

SIZE 3 inches (7½ cm) beak to tail

SHAPE Small, compact hummingbird with short wings, a short tail, and a long, thin beak.

COLOR Green back and vest with off-white underparts. Males have iridescent purple crowns and throats. Females have green heads and white throats.

FEMALE

BEHAVIOR Darts and hovers between desert flowers to sip nectar.

HABITAT Desert scrubland, parks, gardens.

POINT OF FACT Males whistle and show off their purple colors to females while looping and diving around them.

I SAW IT!

WHEN I SAW IT
DATE

WHERE I SAW IT
SPECIFIC LOCATION, INCLUDE STATE

WHAT IT WAS DOING

NOTES

....

....

Greater Roadrunner

(Geococcyx californianus)

Crest on head.

Large long beak.

SIZE 22 inches (56 cm) beak to tail

SHAPE Large, long-legged bird with a long neck and a very long tail.

COLOR Light brown with dark streaks, a black crown with pale spots, and a patch of blue skin behind its eye.

BEHAVIOR Runs quickly on the ground with its neck and tail parallel to the ground.

HABITAT Deserts, scrubland, prairies.

POINT OF FACT Runs as fast as 15 miles (24 km) per hour when chasing prey like scorpions, tarantulas, rodents, lizards, frogs, and birds.

I SAW IT!

WHEN I SAW IT

DATE

WHERE I SAW IT

SPECIFIC LOCATION, INCLUDE STATE

WHAT IT WAS DOING

NOTES

Gila Woodpecker

(Melanerpes uropygialis)

SIZE 9 inches (23 cm) beak to tail

SHAPE Medium-size woodpecker with a long, beefy beak.

COLOR Tan head, neck, and chest, and a black-and-white striped back and wings. Only males have red caps.

BEHAVIOR Clings to and pecks nesting holes in saguaro cacti, calling *churrrr.*

HABITAT Deserts with large cacti, towns, desert washes, and stream sides.

POINT OF FACT Gila is said with the Spanish pronunciation *HEE-luh*. It's the name of a river that flows through New Mexico and Arizona.

I SAW IT!

WHEN I SAW IT

DATE

WHERE I SAW IT

SPECIFIC LOCATION, INCLUDE STATE

WHAT IT WAS DOING

NOTES

Verdin

(Auriparus flaviceps)

SIZE 4 inches (10 cm) beak to tail

SHAPE Small, slim perching bird with a long tail and a small pointy beak.

COLOR Gray body with yellow on its head. Juveniles are all gray.

BEHAVIOR Busy, chatty, nimble bird that flits and forages in thorny brush.

HABITAT Desert scrubland, along desert washes.

POINT OF FACT Builds two kinds of nests: one large, lined, ball-shaped nest with a hole near the bottom for egg laying, and a smaller, lighter, everyday nest for resting and sleeping, or roosting.

YEAR-ROUND

I SAW IT!

WHEN I SAW IT
DATE

WHERE I SAW IT
SPECIFIC LOCATION, INCLUDE STATE

WHAT IT WAS DOING

NOTES

Gambel's Quail

(Callipepla gambelii)

SIZE 9¾ inches (25 cm) beak to tail

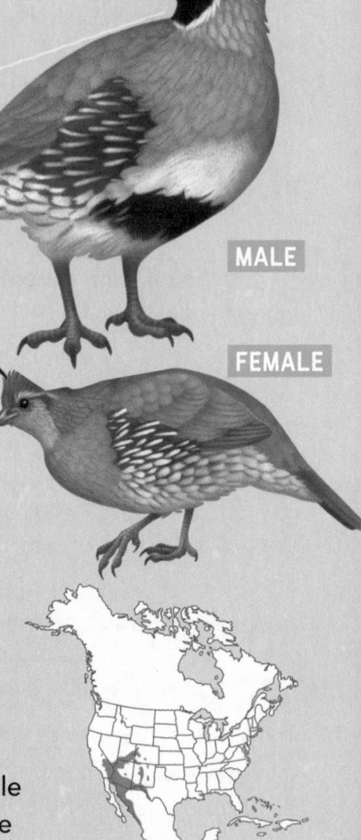

SHAPE Plump, chicken-like bird with a short neck, a small beak, a squared-off tail, and a teardrop-shaped head plume.

COLOR Gray, rust, and cream pattern with streaky sides on its body. Females are grayer with plain heads.

BEHAVIOR Bobs its head plume while walking. In small groups, it scratches for food on the ground.

HABITAT Deserts, mesquite thickets, cactus forests, washes, scrubland.

POINT OF FACT Flocks of quail are called coveys. While foraging for food, one bird acts as the lookout for the covey.

I SAW IT!

WHEN I SAW IT
DATE

WHERE I SAW IT
SPECIFIC LOCATION, INCLUDE STATE

WHAT IT WAS DOING

NOTES

Curve-billed Thrasher

(Toxostoma curvirostre)

Orange eye.

SIZE 10½ inches (26½ cm) beak to tail

Curved beak.

SHAPE Large perching bird with thick legs, a long tail, and a large, curved beak.

COLOR Gray-brown wings, back, and head with a darkly spotted, cream chest.

BEHAVIOR Perches atop spiny cacti while calling out *whit-wheet, whit-wheet.*

HABITAT Deserts, cactus-filled brushlands, canyons, scrubland.

POINT OF FACT Its strong beak is a perfect bug-catching tool because it's able to turn over rocks, flip leaves, and dig into dirt.

YEAR-ROUND

I SAW IT!

WHEN I SAW IT
DATE

WHERE I SAW IT
SPECIFIC LOCATION, INCLUDE STATE

WHAT IT WAS DOING

NOTES

Black-throated Sparrow

(Amphispiza bilineata)

SIZE 5 inches (12½ cm) beak to tail

White eyebrow and chin stripe.

ADULT

Diamond-shaped black throat patch.

JUVENILE

SHAPE Medium-size sparrow with a large head and a chunky bill.

COLOR Gray-brown on its top, a gray face, a creamy belly, and a black throat patch. Juveniles have no throat patch.

BEHAVIOR Pecks for seeds and insects while hopping on the ground.

HABITAT Desert scrubland, washes, and canyons.

POINT OF FACT Females build cup-shaped nests and often line them with animal hair for warmth.

I SAW IT!

WHEN I SAW IT

DATE

WHERE I SAW IT

SPECIFIC LOCATION, INCLUDE STATE

WHAT IT WAS DOING

NOTES

White-throated Sparrow

(Zonotrichia albicollis)

Crown stripes.

Yellow spot between eye and beak.

White throat patch.

ADULT

SIZE 6½ inches (16½ cm) beak to tail

SHAPE Large, plump sparrow with a rounded head, long legs, and a long tail.

COLOR Brown above and gray below with a white throat patch and a black-white-yellow pattern on its face. Juveniles have dull coloring with less yellow.

BEHAVIOR Forages in flocks and scratches leaf litter on the ground to find food.

HABITAT Woodlands, forest and pond edges, old fields, parks, backyards.

POINT OF FACT Whistles its clear but somewhat sad-sounding *Old Sam Peabody, Peabody, Peabody* song all year round.

I SAW IT!

WHEN I SAW IT

DATE

WHERE I SAW IT

SPECIFIC LOCATION, INCLUDE STATE

WHAT IT WAS DOING

NOTES

FOREST BIRDS

Cedar Waxwing

(Bombycilla cedrorum)

White-out-lined black face mask.

Yellow-tipped tail.

SIZE 6 inches (15 cm) beak to tail

ADULT

JUVENILE

SHAPE Medium-size perching bird with a large head, a short neck, a short, squared-off tail, and a wide beak.

COLOR Light brown on its head and chest, gray back and wings, and a light yellow belly. Juveniles are a dull gray and streaky.

BEHAVIOR Flocks gather to gobble fruit from trees and berry bushes while whistling a high *sreee*.

HABITAT Woodlands and forests, farms, orchards, gardens.

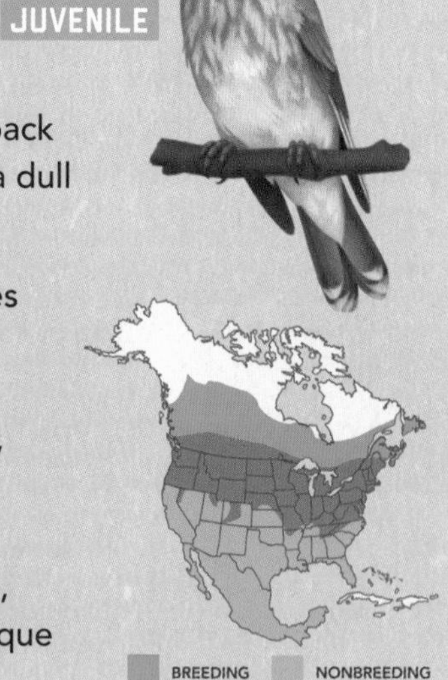

POINT OF FACT Named for the spots on its wings, which look like drops of red wax used to seal antique envelopes.

I SAW IT!

WHEN I SAW IT
DATE

WHERE I SAW IT
SPECIFIC LOCATION, INCLUDE STATE

WHAT IT WAS DOING

NOTES

FOREST BIRDS

Purple Finch

(Haemorhous purpureus)

SIZE 5½ inches (14 cm) beak to tail

SHAPE Large, chunky finch with a large head, a heavy, cone-shaped beak, and a short, notched tail.

COLOR Streaky brown on its cream body. Males have a red wash on their heads and breasts.

BEHAVIOR Sings noisily while searching for food high up in the forest treetops. It flies in an up-and-down wave.

HABITAT Coniferous and deciduous forests, backyards, old fields.

POINT OF FACT The purple in its name comes from the Latin name *purpureus*, which means crimson—a strong red color, not purple!

I SAW IT!

WHEN I SAW IT

DATE

WHERE I SAW IT

SPECIFIC LOCATION, INCLUDE STATE

WHAT IT WAS DOING

NOTES

House Finch

(Haemorhous mexicanus)

MALE

Red rump.

Short, thick beak.

FEMALE

SIZE 5¼ inches (13½ cm) beak to tail

SHAPE Small, slender finch with a flat head, short wings, and a cone-shaped beak.

COLOR Streaky brown back and tail with a red rump, face, and chest. Females are a plain gray-brown with blurry streaks.

BEHAVIOR Flocks perch high in trees and feed on the ground or in plant stalks.

HABITAT City parks, backyards, forest edges.

POINT OF FACT The red coloring of male house finches can vary because it comes from pigments in the food they eat. Its diet while growing a replacement set of feathers, called molting, determines the color of its new feathers.

WHEN I SAW IT

DATE

WHERE I SAW IT

SPECIFIC LOCATION, INCLUDE STATE

WHAT IT WAS DOING

NOTES

Dark-eyed Junco

(Junco hyemalis)

SIZE 6 inches (15 cm) beak to tail

SHAPE Medium-size sparrow with a rounded head, a short, thick beak, and a long tail.

COLOR Varies slightly by region, but it is generally a dark gray with white tail feathers. Females are more gray-brown in color.

BEHAVIOR Flashes its white outer feathers while pumping its tail in flight. Small flocks hop around on the ground to search for seeds.

HABITAT Coniferous or mixed forests, parks, fields, backyards in winter.

POINT OF FACT Also known as the snowbird because it looks like a dark sky over a thin white layer of snow, and also shows up at feeders around first snowfall in some places.

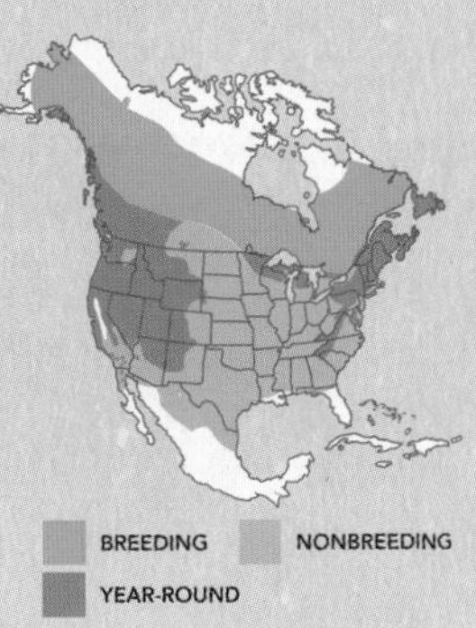

I SAW IT!

WHEN I SAW IT

DATE

WHERE I SAW IT

SPECIFIC LOCATION, INCLUDE STATE

WHAT IT WAS DOING

NOTES

Carolina Wren

(Thryothorus ludovicianus)

Long, white eyebrow stripe.

White chin.

SIZE 5¼ inches (13½ cm) beak to tail

SHAPE Small, round-bodied perching bird with a large head, a small neck, a long tail, and a long, slender beak.

COLOR Light, rusty brown on the top of its belly and a beige-cream color below.

BEHAVIOR Perches and forages in bushes and overgrown vines with its tail bent upward. Its loud call sounds like *teakettle, teakettle*.

HABITAT Woodlands, overgrown fields, bushy thickets, wooded backyards.

POINT OF FACT Wrens are famous for stashing nests in odd places, such as porch light fixtures or old cars.

YEAR-ROUND

I SAW IT!

WHEN I SAW IT

DATE

WHERE I SAW IT

SPECIFIC LOCATION, INCLUDE STATE

WHAT IT WAS DOING

NOTES

FOREST BIRDS

Tufted Titmouse

(Baeolophus bicolor)

Black patch above beak.

Orangish wash on sides.

SIZE 6 inches (15 cm) beak to tail

SHAPE Small, chunky perching bird with a crest on its head, large, round eyes, a thick neck, and a beefy beak.

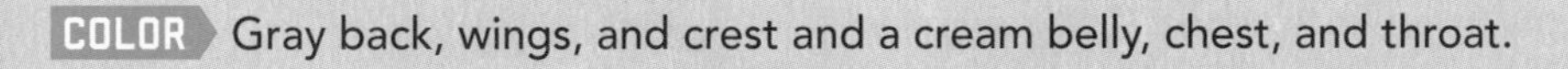

COLOR Gray back, wings, and crest and a cream belly, chest, and throat.

BEHAVIOR Flocks and flies in small groups, often with chickadees, while calling *peter, peter, peter.*

HABITAT Deciduous and mixed coniferous forests, orchards, parks, backyards.

POINT OF FACT Lines its nest with hair or fur, and is known to collect it from living animals—including pets and people.

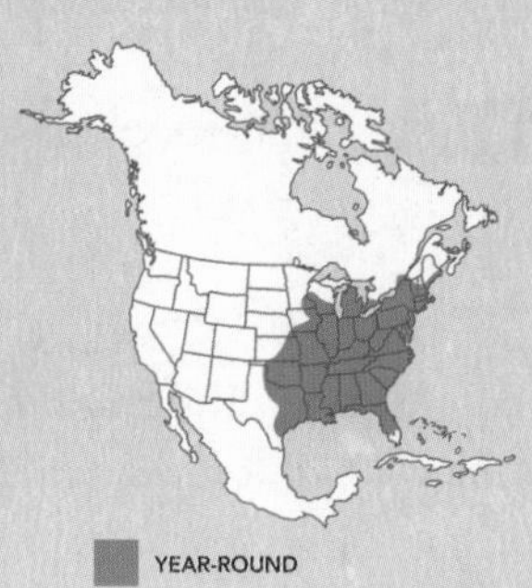

WHEN I SAW IT

DATE

WHERE I SAW IT

SPECIFIC LOCATION, INCLUDE STATE

WHAT IT WAS DOING

NOTES

Red Crossbill

(Loxia curvirostra)

SIZE 6 inches (15 cm) beak to tail

SHAPE Stocky finch with a short, notched tail and a thick, curved beak.

COLOR Red overall with darker brown wings. Females are a dark yellow-green color.

BEHAVIOR Flocks noisily in coniferous forests and clings to cones.

HABITAT Forests of pine, spruce, fir, and other coniferous trees.

POINT OF FACT The size of its beak differs with the cone sizes of the trees where it lives. Larger-beaked birds live in forests with larger cones.

I SAW IT!

WHEN I SAW IT
DATE

WHERE I SAW IT
SPECIFIC LOCATION, INCLUDE STATE

WHAT IT WAS DOING

NOTES

Barred Owl

(Strix varia)

SIZE 19 inches (48½ cm) beak to tail

Smooth round head. Dark brown eyes.

SHAPE Large owl with a round head, no ear tufts, and a rounded tail.

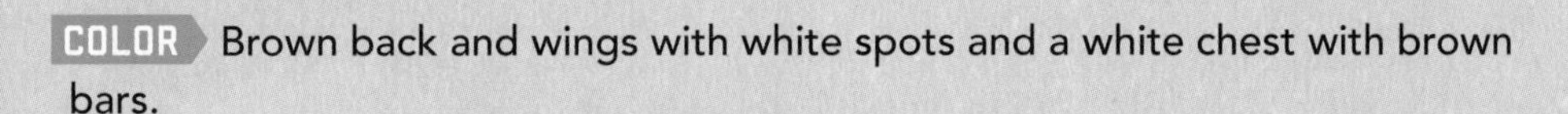

COLOR Brown back and wings with white spots and a white chest with brown bars.

BEHAVIOR Daytime rooster who hunts at night calling *who-cooks-for-you, who-cooks-for-you-all*.

HABITAT Deciduous, coniferous, and swamp forests with large trees.

POINT OF FACT Young barred owls can climb trees! They grab on with their back talons and beak, flap their wings, and shimmy up the tree.

WHEN I SAW IT

DATE

WHERE I SAW IT

SPECIFIC LOCATION, INCLUDE STATE

WHAT IT WAS DOING

NOTES

Great Horned Owl

(Bubo virginianus)

SIZE 23 inches (58½ cm) beak to tail

SHAPE Large, husky owl with a pair of tufts on its head and wide wings.

COLOR Gray-brown, mottled body, a red-brown face, and yellow eyes.

BEHAVIOR Perches at dusk on towers, high fence posts, buildings, and tree branches.

HABITAT Forest edges, swamps, deserts, parks, open woods.

POINT OF FACT The feathery tufts on the top of its head are used to communicate. They also help to camouflage the owl.

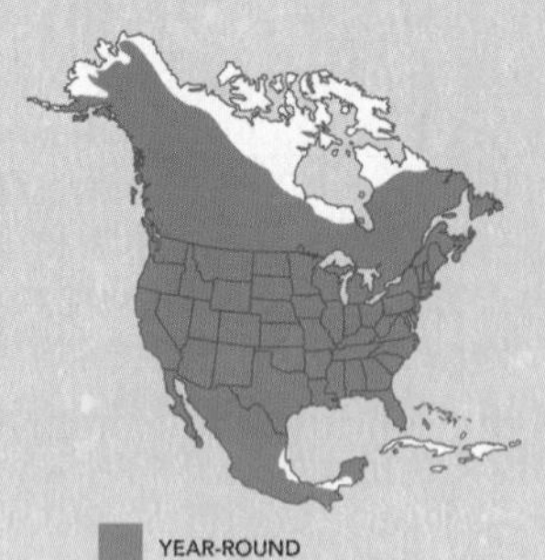

I SAW IT!

WHEN I SAW IT

DATE

WHERE I SAW IT

SPECIFIC LOCATION, INCLUDE STATE

WHAT IT WAS DOING

NOTES

FOREST BIRDS

Black-headed Grosbeak

(Pheucticus melanocephalus)

SIZE 7½ inches (19 cm) beak to tail

SHAPE Large perching bird with a big head and a thick, cone-shaped beak.

COLOR Orange neck and chest, a black head, black-and-white wings, and a gray beak. Females are brown with beige chests.

BEHAVIOR Hops around in treetops while singing.

HABITAT Mixed forests and edges, including in mountains and along desert streams, gardens, backyards.

POINT OF FACT Both males and females sing, sit on eggs, and feed their young.

I SAW IT!

WHEN I SAW IT

DATE

WHERE I SAW IT

SPECIFIC LOCATION, INCLUDE STATE

WHAT IT WAS DOING

NOTES

Scarlet Tanager

(Piranga olivacea)

SIZE 6½ inches (16½ cm) beak to tail

Black wings and tail.

MALE

FEMALE

SHAPE Medium-size perching bird with a large head, a short tail, and a thick, rounded beak.

COLOR Bright red with black wings and a black tail. Females and juveniles are a dull yellow-green with light gray wings.

BEHAVIOR Slowly moves through dense leaves in high treetops to hunt insects.

HABITAT Deciduous and mixed forests of coniferous and deciduous trees.

POINT OF FACT Males are only red while attracting females during the summer breeding season. In the fall, they turn a yellow-green color, but keep their black wings and tail.

I SAW IT!

WHEN I SAW IT

DATE

WHERE I SAW IT

SPECIFIC LOCATION, INCLUDE STATE

WHAT IT WAS DOING

NOTES

FOREST BIRDS

Baltimore Oriole

(Icterus galbula)

SIZE 7 inches (18 cm) beak to tail

SHAPE Medium-size, thick-bodied perching bird with long legs and a pointed beak.

FEMALE

COLOR Orange body, a black head, and black wings. Females are yellow and gray.

BEHAVIOR Sings a slurred, whistle-like song high up in leafy trees while searching for insects, flowers, and fruit.

HABITAT Open woodlands, forest edges, orchards, riverside woods, parks, backyards.

POINT OF FACT It's named for the similarly colored coat-of-arms of Lord Baltimore, the founder of colonial Maryland.

I SAW IT!

WHEN I SAW IT

DATE

WHERE I SAW IT

SPECIFIC LOCATION, INCLUDE STATE

WHAT IT WAS DOING

NOTES

FOREST BIRDS

Black-and-white Warbler

(Mniotilta varia)

SIZE 4¾ inches (12 cm) beak to tail

SHAPE Small, sleek, perching bird (but medium for a warbler!) with a short neck, a flat head, and a long beak.

COLOR Bold black-and-white stripes with more white underneath. Females have more muted stripes with a yellowish tint on their sides.

BEHAVIOR To catch insects, it creeps along branches and dives headfirst on trees.

HABITAT Deciduous and mixed forests of coniferous and deciduous trees.

POINT OF FACT Its habit of probing bark and moss for bugs lives up to its scientific name meaning "moss-plucker."

WHEN I SAW IT

DATE

WHERE I SAW IT

SPECIFIC LOCATION, INCLUDE STATE

WHAT IT WAS DOING

NOTES

Yellow-rumped Warbler

(Setophaga coronata)

SIZE 5 inches (12½ cm) beak to tail

SHAPE Small perching bird (but large for a warbler!) with a big head and a long, thin tail.

COLOR In summer, males are gray with yellow on their faces, sides, and rumps. Females are dull with some brown. All winter birds are pale and brown.

BEHAVIOR Busily flies from tree to tree while flashing its yellow rump.

HABITAT Open coniferous forests and edges in summer. Woodlands, parks, backyards in winter.

POINT OF FACT Also known as the myrtle warbler, as it's the only warbler able to digest waxy bayberries and wax myrtle seeds. This helps it survive farther north in winter than other warblers.

I SAW IT!

WHEN I SAW IT
DATE

WHERE I SAW IT
SPECIFIC LOCATION, INCLUDE STATE

WHAT IT WAS DOING

NOTES

....

....

Ovenbird

(Seiurus aurocapilla)

SIZE 5¼ inches (13½ cm) beak to tail

SHAPE Small, chunky perching bird (but large for a warbler!) with a round head and a thick beak.

COLOR Olive-green back and wings, a cream chest with black streaky spots, and pink legs.

BEHAVIOR Forages on the ground with its tail up and loudly sings *tea-CHER, tea-CHER, tea-CHER.*

HABITAT Deciduous and mixed deciduous coniferous forests.

POINT OF FACT Its name comes from its dome-shaped nest, which looks like an old-fashioned, wood-fired oven.

I SAW IT!

WHEN I SAW IT
DATE

WHERE I SAW IT
SPECIFIC LOCATION, INCLUDE STATE

WHAT IT WAS DOING

NOTES

FOREST BIRDS

Eastern Wood-pewee

(Contopus virens)

SIZE 6 inches (15 cm) beak to tail

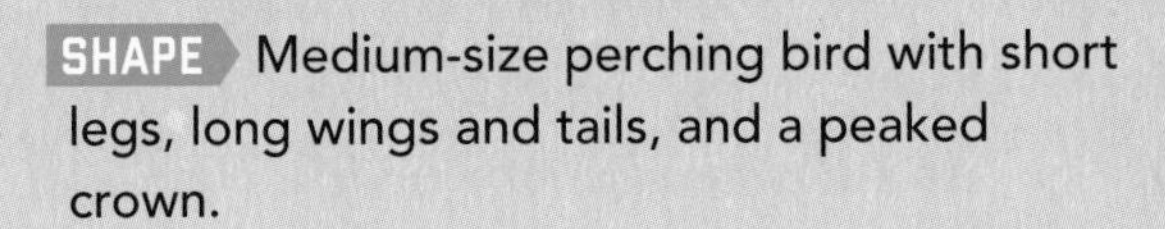

SHAPE Medium-size perching bird with short legs, long wings and tails, and a peaked crown.

COLOR Brown-gray back and dark wings, a white chest and belly, and a yellow lower beak.

BEHAVIOR Flicks its wings as it lands in high trees, then perches upright and whistles *PEE-we-EE.*

HABITAT Deciduous forests, smaller woodlots.

POINT OF FACT Covers its nest with lichen to camouflage it from blue jays, snakes, squirrels, and other predators.

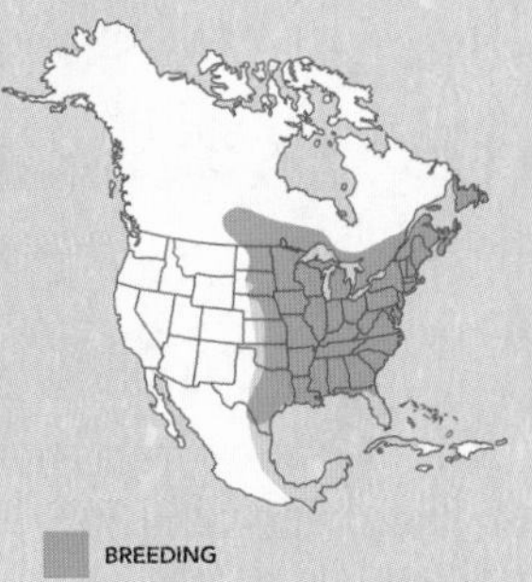

I SAW IT!

WHEN I SAW IT

DATE

WHERE I SAW IT

SPECIFIC LOCATION, INCLUDE STATE

WHAT IT WAS DOING

NOTES

Hermit Thrush

(Catharus guttatus)

SIZE 6¼ inches (16 cm) beak to tail

SHAPE Medium-size perching bird with a round head, a straight beak, and a long tail.

COLOR Brown head, back, and wings, a rusty-red tail, and a creamy, white-spotted throat and chest.

BEHAVIOR Hops through leaf litter, and raises and lowers its tail when perching.

HABITAT Forest understories, woodland edges and openings.

POINT OF FACT It has a gorgeous, flute-like song that starts low and soars upward in pitch like its cousin, the robin. All are famously great songsters!

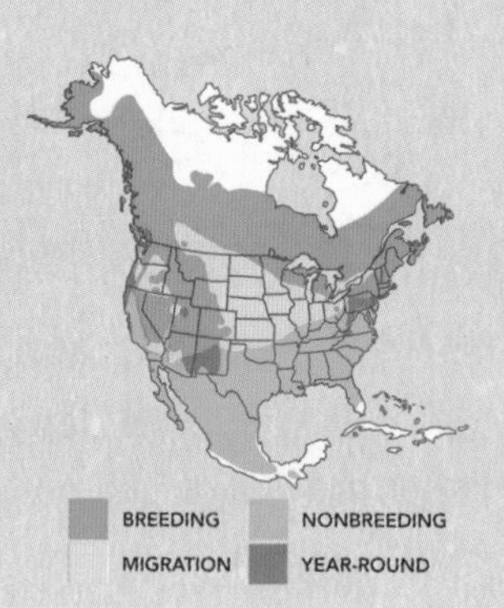

I SAW IT!

WHEN I SAW IT

DATE

WHERE I SAW IT

SPECIFIC LOCATION, INCLUDE STATE

WHAT IT WAS DOING

NOTES

....

....

FOREST BIRDS

Pileated Woodpecker

(Dryocopus pileatus)

Beak nearly as long as head is.

Red cheek stripe on males only.

SIZE 17 inches (43 cm) beak to tail

SHAPE Huge woodpecker with a long neck, a pointy head crest, and a long, chiseled beak.

COLOR Black with white stripes on its face and neck, and a bright red head crest. Males have a red cheek stripe.

BEHAVIOR Clings to dead trees while drilling out rectangular holes, and flies in an up-and-down pattern.

HABITAT Forests and other areas with standing dead trees, including backyards and parks.

POINT OF FACT Its call is long and loud, and sounds a bit like sped-up, crazy laughter.

YEAR-ROUND

I SAW IT!

WHEN I SAW IT

DATE

WHERE I SAW IT

SPECIFIC LOCATION, INCLUDE STATE

WHAT IT WAS DOING

NOTES

Yellow-bellied Sapsucker

(Sphyrapicus varius)

MALE

Red throat and crown.

Long, vertical white wing slash.

FEMALE

SIZE 8 inches (20½ cm) beak to tail

SHAPE Small woodpecker with a thick, straight, wedge-like beak.

COLOR Black-and-white back and wings, a cream to pale yellow belly, and a red crown. Males also have red throats.

BEHAVIOR Clings upright on tree trunks to feed at rows of holes drilled into bark.

HABITAT Deciduous and coniferous forests.

POINT OF FACT It drills feeding holes called sapwells to eat both the sugary tree sap and any insects stuck in the sap!

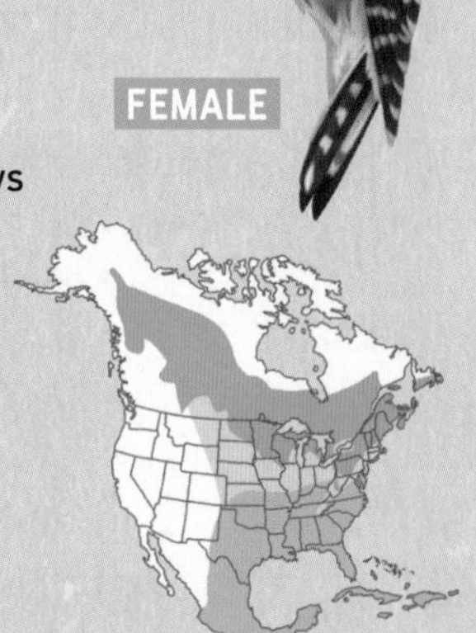

I SAW IT!

WHEN I SAW IT

DATE

WHERE I SAW IT

SPECIFIC LOCATION, INCLUDE STATE

WHAT IT WAS DOING

NOTES

FOREST BIRDS

Downy Woodpecker

(Dryobates pubescens)

SIZE 6 inches (15 cm) beak to tail

White bands on black wings.

Beak is half as wide as head.

MALE

FEMALE

SHAPE Small woodpecker with a square-ish head and a short, chiseled beak.

COLOR Black tail, black wings with white bands, a white belly, and a black-and-white striped head. Males have red spots on the back of their heads.

BEHAVIOR Clings and scoots up and down trees, pecking at the bark. It flies in an up-and-down wave pattern.

HABITAT Open deciduous woodlands, forest edges, parks, backyards.

POINT OF FACT It's noisy in the spring and summer! It drums on hollow branches and makes a shrill, rattling *trrrrrrrrrr* call.

YEAR-ROUND

I SAW IT!

WHEN I SAW IT

DATE

WHERE I SAW IT

SPECIFIC LOCATION, INCLUDE STATE

WHAT IT WAS DOING

NOTES

Wood Duck

(Aix sponsa)

SIZE 19 inches (48½ cm) beak to tail

SHAPE Medium-size duck with a thin neck, a thick tail, short wings, a large head, and a slicked-back crest.

COLOR Shiny green head with white stripes, a red-brown chest, red eyes, and a colorful beak. Females are dull brown with white eye rings.

BEHAVIOR Jerks its head back and forth when swimming. Small flocks whistle *uh-wheek* in flight.

HABITAT Wooded swamps, forests near small lakes and ponds, marshes.

POINT OF FACT It's one of the few ducks with strong, bark-gripping claws on their webbed feet. It can perch on branches and nest in tree cavities.

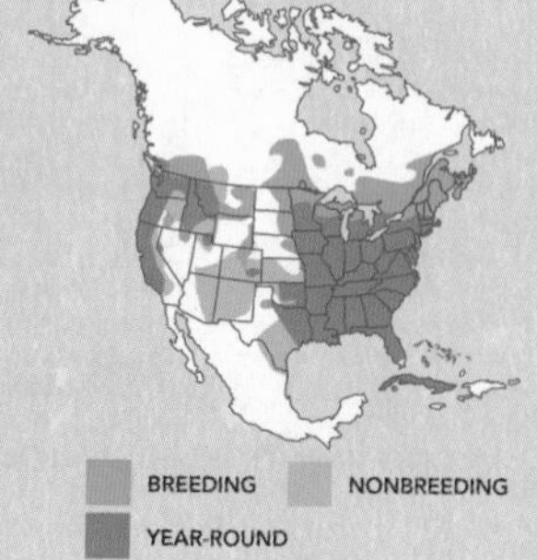

I SAW IT!

WHEN I SAW IT
DATE

WHERE I SAW IT
SPECIFIC LOCATION, INCLUDE STATE

WHAT IT WAS DOING

NOTES

....

....

Wild Turkey

(Meleagris gallopavo)

SIZE 44 inches (112 cm) beak to tail

SHAPE Big, plump game bird with long legs, a small head, a long neck, and a wide, rounded tail. Females have smaller tails.

COLOR Dark brown with bars on its wings and a gray-to-red head and neck skin.

BEHAVIOR Walks along the ground in flocks to scratch for food. When courting, males puff up, display a fan of tail feathers, and gobble.

HABITAT Woods of mature oak, hickory, and beech trees as well as fields and forest edges.

POINT OF FACT Flies quickly for short distances, sprinting up to 60 miles (96 km) per hour.

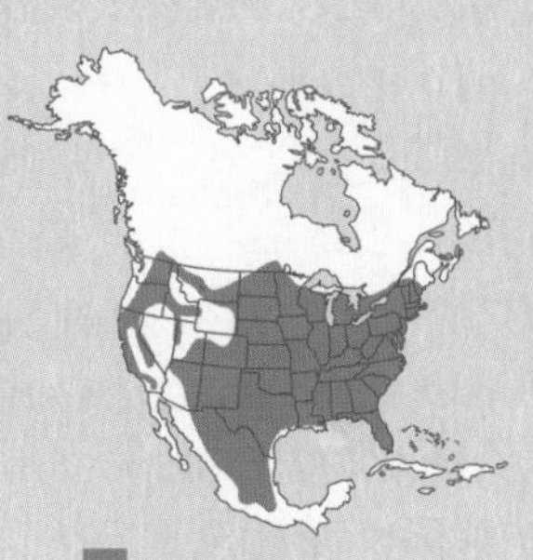

WHEN I SAW IT
DATE

WHERE I SAW IT
SPECIFIC LOCATION, INCLUDE STATE

WHAT IT WAS DOING

NOTES

Red-shouldered Hawk

(Buteo lineatus and Buteo elegans.)

SIZE 18 inches (45½ cm) beak to tail

SHAPE Medium-size hawk with wide, rounded wings and a medium-length tail.

COLOR Rusty shoulder patches, brown-and-white wings, and a black tail with white bands. Juveniles are brown on top and white with brown streaks underneath.

JUVENILE

BEHAVIOR Hunts from high perches and while soaring over forests. Its call is a high, rising *keee-yah* scream.

HABITAT Deciduous forests, fields near woodlands.

POINT OF FACT In flight, look for a swath of bright feathers toward the end of the hawk's wing tips. They're called "wing windows" because light appears to shine through.

BREEDING NONBREEDING
YEAR-ROUND

I SAW IT!

WHEN I SAW IT
DATE

WHERE I SAW IT
SPECIFIC LOCATION, INCLUDE STATE

WHAT IT WAS DOING

NOTES

Belted Kingfisher

(Megaceryle alcyon)

Long beak.

MALE

Crest on head.

FEMALE

SIZE 12 inches (30½ cm) beak to tail

SHAPE Big-headed, husky bird with a shaggy crest and a long, thick beak.

COLOR Blue-gray head, wings, and chest with a white belly and neck. Females have rusty bands on their bellies.

BEHAVIOR Flies over the water while making a loud, rattling, scolding call.

HABITAT Near streams, rivers, ponds, lakes, and estuaries.

POINT OF FACT Dig tunnels into riverbanks in order to nest.

I SAW IT!

WHEN I SAW IT
DATE

WHERE I SAW IT
SPECIFIC LOCATION, INCLUDE STATE

WHAT IT WAS DOING

NOTES

....

....

Great Blue Heron

(Ardea herodias)

SIZE 45 inches (114½ cm) beak to tail

SHAPE Large, tall water bird with long legs, a long neck, and a big, thick beak.

COLOR Light blue-gray with black on its head, a yellow beak, and a black head plume. Juveniles are dull and streaky.

BEHAVIOR Wades in shallow water while holding its neck in an S shape to stalk prey. It flies with painfully slow wingbeats, a scrunched-up neck, and trailing legs.

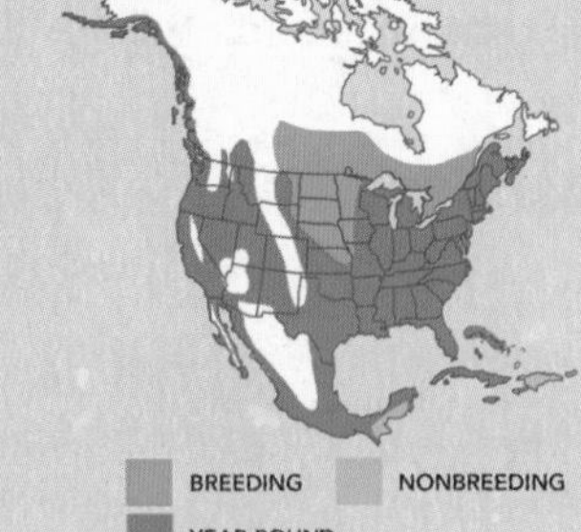

HABITAT Both freshwater and saltwater marshes and coasts, riverbanks, lakes, and ponds.

POINT OF FACT Like many herons, it lives in large, noisy colonies (called rookeries) in big trees filled with sloppy nests.

I SAW IT!

WHEN I SAW IT
DATE

WHERE I SAW IT
SPECIFIC LOCATION, INCLUDE STATE

WHAT IT WAS DOING

NOTES

....

....

American White Pelican

(Pelecanus erythrorhynchos)

SIZE 60 inches (152½ cm) beak to tail

Huge, yellow beak.

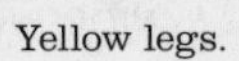

SHAPE Huge pelican with a thick body, an enormous beak, wide wings, a long neck, short legs, and a square tail.

COLOR White body, yellow legs, and a yellow beak. Black feathers can only be seen when it's flying.

BEHAVIOR Flocks forage on the surface of lakes by tipping up their beaks to scoop prey out of the water.

HABITAT Lakes and marshes, coasts, bays, estuaries.

POINT OF FACT Its large expandable bill pouch can hold more than 2½ gallons (9½ liters) of fish-filled water.

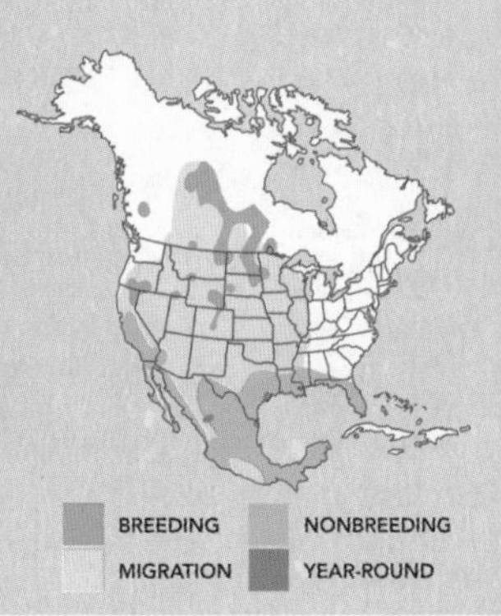

I SAW IT!

WHEN I SAW IT

DATE

WHERE I SAW IT

SPECIFIC LOCATION, INCLUDE STATE

WHAT IT WAS DOING

NOTES

....

....

Bald Eagle

(Haliaeetus leucocephalus)

SIZE 32 inches (81½ cm) beak to tail

JUVENILE

SHAPE Huge, beefy raptor with a long, hooked beak and a wide wingspan.

COLOR White head, dark brown body and wings, and yellow legs. Juveniles are all brown with white flecks and splotches.

BEHAVIOR Perches in trees, on the ground, or soars high overhead. It flaps slowly when flying tree to tree.

HABITAT Near lakes, rivers, marshes, and coasts.

POINT OF FACT It is notorious for stealing fish from the mouths of osprey and other birds smaller than itself.

I SAW IT!

WHEN I SAW IT

DATE

WHERE I SAW IT

SPECIFIC LOCATION, INCLUDE STATE

WHAT IT WAS DOING

NOTES

....

....

Double-crested Cormorant

(Phalacrocorax auritus)

Orange face patch.

Hooked beak.

SIZE 32 inches (81½ cm) beak to tail

SHAPE Large waterbird with a long, twisted neck, a small head, and a hooked beak.

COLOR Black-brown with orange skin on its face. Juveniles are a lighter brown with pale chests and necks.

BEHAVIOR Floats low in the water, showing just an oval of its back behind its long, erect neck. It perches in the sun with its wings spread wide to dry.

HABITAT Fish-filled fresh and salt bodies of water including lakes, coasts, rivers, and large ponds.

POINT OF FACT It searches for fish as it floats. When its prey is spotted, it dives down as deep as 24½ feet (7½ m) under the water to chase fish.

I SAW IT!

WHEN I SAW IT

DATE

WHERE I SAW IT

SPECIFIC LOCATION, INCLUDE STATE

WHAT IT WAS DOING

NOTES

Canada Goose

(Branta canadensis)

SIZE 35 inches (89 cm) beak to tail

SHAPE Large goose with a long neck, big, webbed feet, and a wide beak.

COLOR Tan to brown body, a black neck, and a black head with white cheeks.

BEHAVIOR Flocks fly in a V shape and graze on large lawns, fields, and open areas. It floats high in the water and pokes its head and neck under the surface to forage.

HABITAT Near lakes, rivers, ponds, parks, farm fields.

POINT OF FACT Even before hatching, chicks inside the eggs peep and trill to communicate with their parents.

BREEDING
NONBREEDING
YEAR-ROUND

I SAW IT!

WHEN I SAW IT

DATE

WHERE I SAW IT

SPECIFIC LOCATION, INCLUDE STATE

WHAT IT WAS DOING

NOTES

Tree Swallow

(Tachycineta bicolor)

SIZE 5½ inches (14 cm) beak to tail

Tail no longer than wings.

Black eye mask.

MALE

FEMALE

SHAPE Slender, sleek, small perching bird with long, pointed wings and a short tail.

COLOR Iridescent blue-green on top and white below. Females are dull and brown on top.

BEHAVIOR Small flocks swoop and dive through the air to catch flying insects.

HABITAT Marshes, wooded swamps, ponds, fields, parks.

POINT OF FACT Nests in old trees with holes that have been made by woodpeckers.

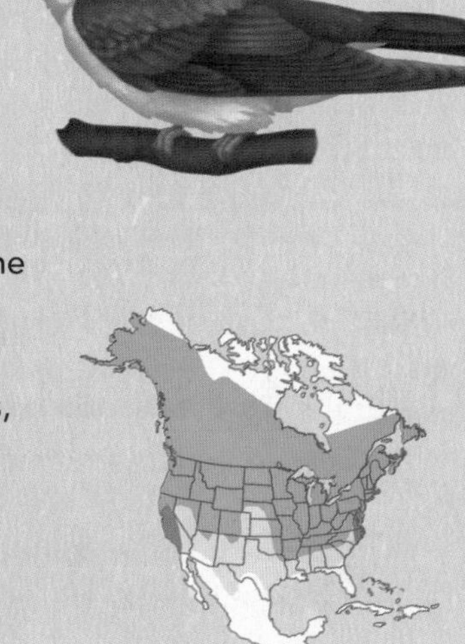

I SAW IT!

WHEN I SAW IT

DATE

WHERE I SAW IT

SPECIFIC LOCATION, INCLUDE STATE

WHAT IT WAS DOING

NOTES

....

....

Snowy Egret

(Egretta thula)

SIZE 24 inches (61 cm) beak to tail

SHAPE Medium-size heron with a long neck, a small head, and long, skinny legs.

COLOR All white with a black beak and legs, and yellow feet. Juveniles have green legs.

BEHAVIOR Wades in shallow water, hunting patiently for fish to spear.

HABITAT Fresh, salt, and brackish water coasts, shorelines, mudflats, and marshes.

POINT OF FACT It sometimes pats the surface of the water with its foot to attract fish to the top.

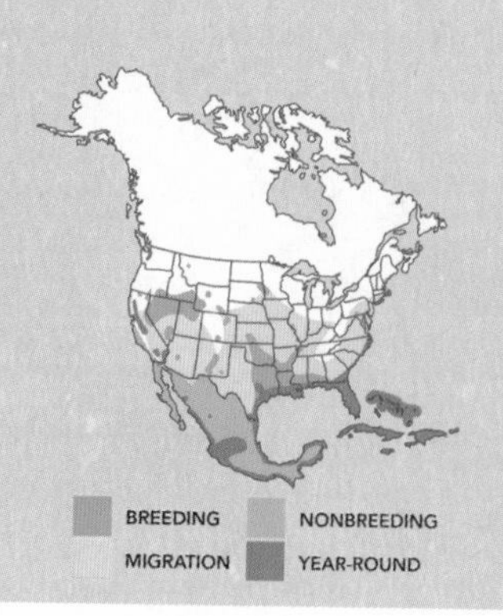

I SAW IT!

WHEN I SAW IT

DATE

WHERE I SAW IT

SPECIFIC LOCATION, INCLUDE STATE

WHAT IT WAS DOING

NOTES

....

....

BIRDS AROUND WATER

Osprey

(Pandion haliaetus)

SIZE 22 inches (56 cm) beak to tail

SHAPE Large hawk with narrow wings and long legs.

COLOR Brown back, tail, and wing tops with a white chest, white legs, and a white belly. Its head is white with a brown stripe. Juveniles are white-spotted on their backs with cream breasts.

BEHAVIOR Flies with a shrugged-shoulder posture that puts a crook in its wings.

HABITAT Rivers, ponds, lakes, coasts.

POINT OF FACT Also called the "fish hawk," it catches fish with its talons by diving feetfirst into the water. On average, it catches a fish after only 12 minutes of hunting.

I SAW IT!

WHEN I SAW IT

DATE

WHERE I SAW IT

SPECIFIC LOCATION, INCLUDE STATE

WHAT IT WAS DOING

NOTES

....

....

Sandhill Crane

(Antigone canadensis)

SIZE 47 inches (119½ cm) beak to tail

SHAPE Tall, big bird with a long neck, long legs, and a long beak.

COLOR Gray body with some rusty brown in the summer, a red crown, and white cheeks. Juveniles have no red crowns.

BEHAVIOR Flocks walk along in grasslands to forage for grains and wade into wetlands to hunt for invertebrates.

HABITAT Open grasslands and wetlands, farm fields.

POINT OF FACT It dances when it's courting by pumping its head up and down, bowing, and leaping with its wings outstretched. It's quite a sight!

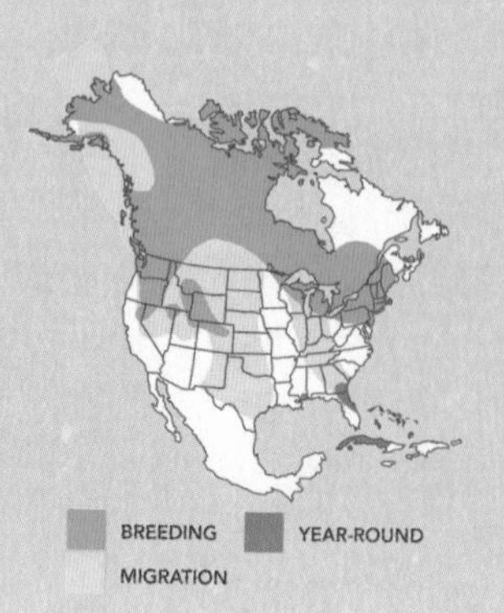

I SAW IT!

WHEN I SAW IT
DATE

WHERE I SAW IT
SPECIFIC LOCATION, INCLUDE STATE

WHAT IT WAS DOING

NOTES

American Coot

(Fulica americana)

SIZE 16 inches (40½ cm) beak to tail

SHAPE Football-shaped, chicken-like bird with short wings, large feet, and a sloping beak.

COLOR Gray-black with a white beak and a white forehead. Juveniles are pale gray with darker beaks. Just-hatched chicks are covered in a rusty-red down, have bare red spots on their heads, and have red beaks.

BEHAVIOR Bobs its head forward and backward while swimming, noisily cackling and chattering. It walks like a chicken on land.

HABITAT Park ponds, marshes, lakes, and salt marshes.

POINT OF FACT Chick looks a bit like a bald man in a turtleneck sweater.

BREEDING
NONBREEDING
MIGRATION
YEAR-ROUND

I SAW IT!

WHEN I SAW IT

DATE

WHERE I SAW IT

SPECIFIC LOCATION, INCLUDE STATE

WHAT IT WAS DOING

NOTES

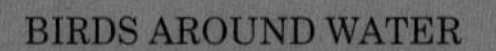

Black Tern

(Chlidonias niger)

SIZE 9½ inches (24 cm) beak to tail

SHAPE Small tern with a forked tail, pointy wings, and a sharp beak.

COLOR Gray back and wings with a black head, chest, and belly. Nonbreeding (fall and winter) adults have mostly white heads and chests.

BEHAVIOR Large flocks fly in crazy patterns to catch insects out of the air.

HABITAT Freshwater marshes, lakes.

POINT OF FACT Breeding pairs perform daredevil courtship flights, zooming up to 600 feet (183 m) into the air and then chasing each other with swoops and dives while calling loudly for as long as 20 minutes.

I SAW IT!

WHEN I SAW IT
DATE

WHERE I SAW IT
SPECIFIC LOCATION, INCLUDE STATE

WHAT IT WAS DOING

NOTES

Lesser Yellowlegs

(Tringa flavipes)

SIZE 9½ inches (24 cm) beak to tail

SHAPE Small, slender shorebird with a small head, long legs, and a thin beak.

COLOR Dusk-brown and gray with flecks of white on its back, wings, and head. It has a lighter-colored chest and belly.

BEHAVIOR Walks in shallow water to probe for prey under the surface.

HABITAT Shallow fresh and saltwater marshes, mudflats, and streams.

POINT OF FACT Females leave the breeding area first, leaving the fathers to finish raising the young.

I SAW IT!

WHEN I SAW IT

DATE

WHERE I SAW IT

SPECIFIC LOCATION, INCLUDE STATE

WHAT IT WAS DOING

NOTES

Green Heron

(Butorides virescens)

SIZE 17 inches (43 cm) beak to tail

SHAPE Small, short, stocky heron with a thick neck, rounded wings, and a long beak.

COLOR Dark green on its back, wings, and head with a dark, red-brown chest and neck. Juveniles have necks that are brown with light streaks and spotted wings.

BEHAVIOR Perches with its neck hunched up on low branches or logs over the water. It flies with a crooked neck while squawking loudly.

HABITAT Wooded ponds, marshes, rivers, lakes, and estuaries.

POINT OF FACT Its daggerlike beak is perfect for spearing fish. Watch out!

I SAW IT!

WHEN I SAW IT

DATE

WHERE I SAW IT

SPECIFIC LOCATION, INCLUDE STATE

WHAT IT WAS DOING

NOTES

....

....

Prothonotary Warbler

(Protonotaria citrea)

SIZE 5 inches (12½ cm) beak to tail

SHAPE Small, round perching bird (but large for a warbler!) with a thick neck, a large head, short legs, a short tail, and a medium-length beak.

COLOR Bright yellow head and chest, blue-gray wings and tail, a green back, and white underneath its tail. Females are dull with green on their crowns and napes.

BEHAVIOR Hops patiently along fallen logs, low branches, or on the ground to forage near slow water.

HABITAT Wooded swamps, flooded forests, and other areas near streams and lakes.

POINT OF FACT Gets its name from the bright yellow robes once worn by a group of Catholic church scribes known as prothonotaries.

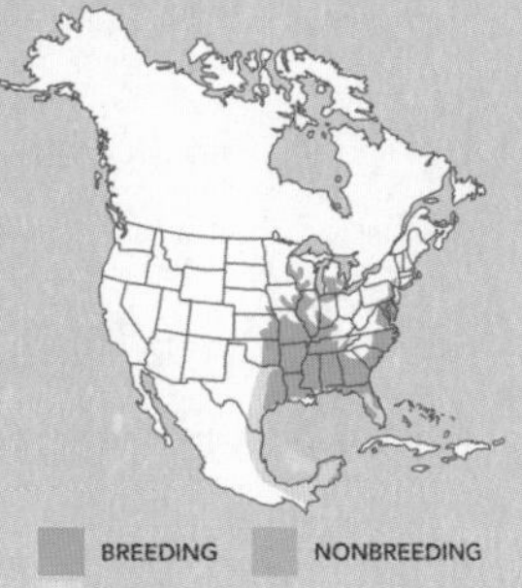

I SAW IT!

WHEN I SAW IT
DATE

WHERE I SAW IT
SPECIFIC LOCATION, INCLUDE STATE

WHAT IT WAS DOING

NOTES

Yellow-headed Blackbird

(Xanthocephalus xanthocephalus)

SIZE 9 inches (23 cm) beak to tail

White wing patch.

MALE

Large beak.

SHAPE Beefy, large perching bird with a big head and a cone-shaped beak.

COLOR Glossy black with a yellow head and a yellow chest. Females and juveniles are brown and a dull yellow.

BEHAVIOR Males sing straight up into the sky from atop cattails while showing off their colors. Their heads roll back and tilt to the side.

HABITAT Freshwater wetlands with dense plants in summer. Farm fields in winter.

POINT OF FACT The male's song is harsh and hoarse. It sounds a bit like someone retching!

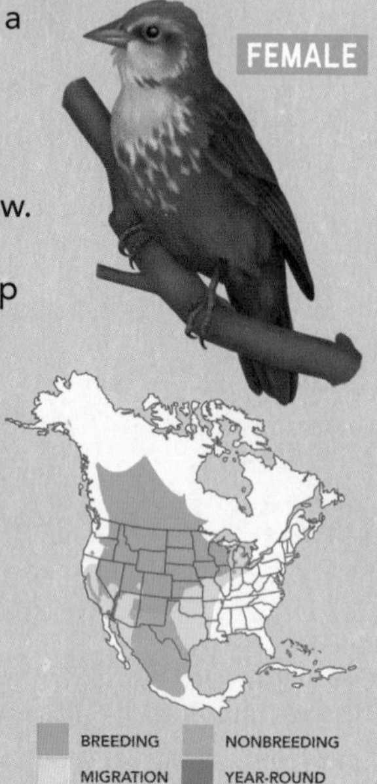

I SAW IT!

WHEN I SAW IT

DATE

WHERE I SAW IT

SPECIFIC LOCATION, INCLUDE STATE

WHAT IT WAS DOING

NOTES

....

....

Common Yellowthroat

(Geothlypis trichas)

MALE

Black mask. White line over mask. Yellow throat.

SIZE 4¾ inches (12 cm) beak to tail

SHAPE Small, round perching bird (and small for a warbler) with a thick neck.

COLOR Green-brown back, wings, and crown with a bright yellow throat and black mask. Females are green-brown with dull yellow throats and no masks.

BEHAVIOR Lurks in dense, bushy thickets to search for bugs. It loudly sings a clear, ringing *witchety-witchety-witchety.*

HABITAT Marsh, grassland, and forest edges with thick, low vegetation.

POINT OF FACT One of its folk names, the black-masked ground warbler, describes it well.

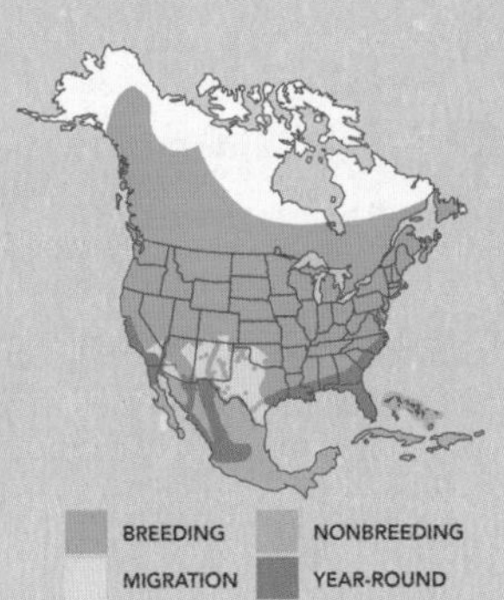

I SAW IT!

WHEN I SAW IT

DATE

WHERE I SAW IT

SPECIFIC LOCATION, INCLUDE STATE

WHAT IT WAS DOING

NOTES

Barn Swallow

(Hirundo rustica)

Long outer tail feathers.

SIZE 6¾ inches (17 cm) beak to tail

Rusty face.

SHAPE Sparrow-size perching bird with a flat head, wide shoulders, long, pointy wings, and a forked tail.

COLOR Shiny, dark-blue back, wings, and tail with a rusty orange face, throat, chest, and belly.

BEHAVIOR Flies with quick turns and fast flapping to hunt and munch on insects in the air.

HABITAT Marshes, meadows, fields, parks, and other open areas.

POINT OF FACT How did it get its forked tail? Legend says it stole fire from the gods who angrily hurled a burning ember at it, singeing away its middle tail feathers.

I SAW IT!

WHEN I SAW IT

DATE

WHERE I SAW IT

SPECIFIC LOCATION, INCLUDE STATE

WHAT IT WAS DOING

NOTES

....

....

OPEN COUNTRY BIRDS

Bobolink

(Dolichonyx oryzivorus)

MALE

Short, pointy black beak.

White back.

SIZE 7 inches (18 cm) beak to tail

SHAPE Small perching bird with a large head, a short neck, and a short tail.

COLOR Black with a white rump and back, and a light, orange-tan nape. Females and wintertime males are streaky brown with brown stripes on their crowns, with pink beaks.

BEHAVIOR Sings, clings to stalks of tall grass full of seeds, and flutters its wings to fly low.

HABITAT Tall grasslands, overgrown fields, prairies.

POINT OF FACT Its song is a series of buzzes and whistles that sound somewhat electronic, almost like R2-D2 from *Star Wars*.

BREEDING
MIGRATION

I SAW IT!

WHEN I SAW IT

DATE

WHERE I SAW IT

SPECIFIC LOCATION, INCLUDE STATE

WHAT IT WAS DOING

NOTES

....

....

Dickcissel

(Spiza americana)

SIZE 6 inches (15 cm) beak to tail

Rusty shoulder patch.

MALE

Black throat V.

FEMALE

SHAPE Small, chesty perching bird with a large, thick beak and a short tail.

COLOR Brown and gray back with rusty shoulders, a gray head, a yellow face and chest, a black throat, and a gray beak. Females are dull with no black throats.

BEHAVIOR Perches on shrubs, fences, and tall stalks while singing its name: *dick-dick-dick-ciss-ciss-ell.*

HABITAT Tall grasslands, prairies, hayfields, old pastures, roadsides.

POINT OF FACT Its strong beak can hull and eat a dozen seeds per minute. Males win the eating contest, shelling seeds a bit faster than females.

BREEDING
MIGRATION

WHEN I SAW IT
DATE

WHERE I SAW IT
SPECIFIC LOCATION, INCLUDE STATE

WHAT IT WAS DOING

NOTES

....

....

Western Meadowlark

(Sturnella neglecta)

BREEDING

SIZE 8½ inches (21½ cm) beak to tail

Long, slender beak.

V on chest.

NON-BREEDING

SHAPE Medium-size, husky perching bird with a slender beak, a flat head, a short neck, rounded wings, and a short, stiff tail.

COLOR Brown-and-cream patterned wings, back, and tail with a bright yellow throat and chest. A black V fades outside its breeding season.

BEHAVIOR Sits atop fence posts, power lines, and bushes while belting out a flute-like song.

HABITAT Open grasslands and farm fields, marsh edges, roadsides, mountain meadows.

POINT OF FACT It probes the soil and manure piles for beetles, ants, grasshoppers, and worms.

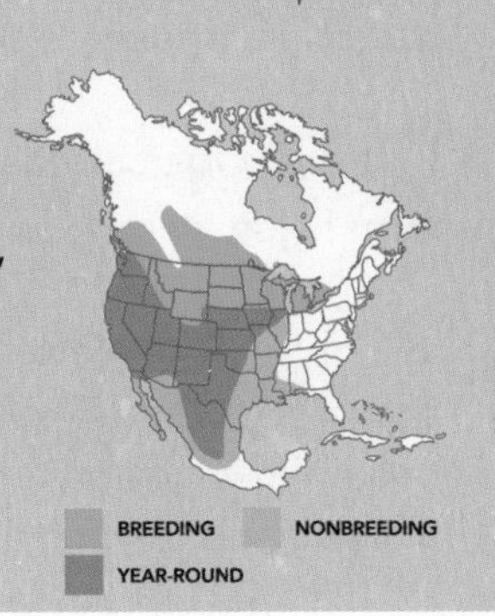

I SAW IT!

WHEN I SAW IT

DATE

WHERE I SAW IT

SPECIFIC LOCATION, INCLUDE STATE

WHAT IT WAS DOING

NOTES

....

....

Horned Lark

(Eremophila alpestris)

SIZE 7 inches (18 cm) beak to tail

SHAPE Small perching bird with a short beak, a round head, and a set of horn-like feathers on the back of its head.

FEMALE

COLOR Sand to rusty brown on its back, wings, and head, with a yellow throat, a cream belly, and a black face mask and collar. Females are dull with no face masks.

BEHAVIOR Flocks forage on the bare soil and ground for seeds and insects. It sings a high, tinkling-glass–like song.

HABITAT Open areas with short grass or bare ground, farm fields, desert, scrubland, tundra.

POINT OF FACT Males fiercely defend their territory during breeding season. Rival males battle it out, punching with extended wings on the ground and flying at each other, pecking and clawing as they rise into the air.

I SAW IT!

WHEN I SAW IT
DATE

WHERE I SAW IT
SPECIFIC LOCATION, INCLUDE STATE

WHAT IT WAS DOING

NOTES

....

....

Northern Harrier

(Circus hudsonius)

SIZE 18½ inches (47 cm) beak to tail

SHAPE Medium-size hawk with a wide wingspan, a long, rounded tail, and an owl-like face.

Small, sharp, hooked beak. Round, flat face.

COLOR Males are gray on top and white underneath with black wing tips. Females and juveniles are brown with streaks on their chests and bellies. All of them have white rump patches.

BEHAVIOR Flies low over fields and marshes to hunt mice and small birds.

HABITAT Open grasslands, prairies, fields, and marshes.

POINT OF FACT It hunts by both sight and sound. A disc-shaped face helps funnel faint sounds toward its ears, just like owls!

BREEDING NONBREEDING YEAR-ROUND

I SAW IT!

WHEN I SAW IT

DATE

WHERE I SAW IT

SPECIFIC LOCATION, INCLUDE STATE

WHAT IT WAS DOING

NOTES

....

....

OPEN COUNTRY BIRDS

Turkey Vulture

(Cathartes aura)

SIZE 27 inches (68½ cm) beak to tail

SHAPE Huge raptor with long, wide wings that end in fingerlike tips, a long tail, a tiny head, and a big beak.

COLOR Dark brown with a naked red head and a cream beak. The undersides of its tail and wings are light gray in flight. Legs are red but often covered in vulture's own chalky white poop.

BEHAVIOR Teeters unsteadily as it soars with its wings raised into a very shallow V shape.

HABITAT Farm fields, suburbs, roadsides, open country.

POINT OF FACT It doesn't sing or call, but it does hiss and groan when predators approach.

I SAW IT!

WHEN I SAW IT

DATE

WHERE I SAW IT

SPECIFIC LOCATION, INCLUDE STATE

WHAT IT WAS DOING

NOTES

....

....

OPEN COUNTRY BIRDS

Killdeer

(Charadrius vociferus)

SIZE 9½ inches (24 cm) beak to tail

SHAPE Medium shorebird with a large head, big eyes, a short beak, long wings and legs, and a pointy tail.

COLOR Light brown on its back and wings with a white chest and belly. It also has black bands on its chest and face, and an orange rump that's visible when it's flying.

BEHAVIOR Runs along the ground, then stops, looks, then runs again. It flies overhead while scolding *tee-deee, tee-deee*.

HABITAT Open areas with patches of bare ground like plowed fields, sandbars, scrubby pastures, mudflats, parking lots, lawns, driveways, and golf courses.

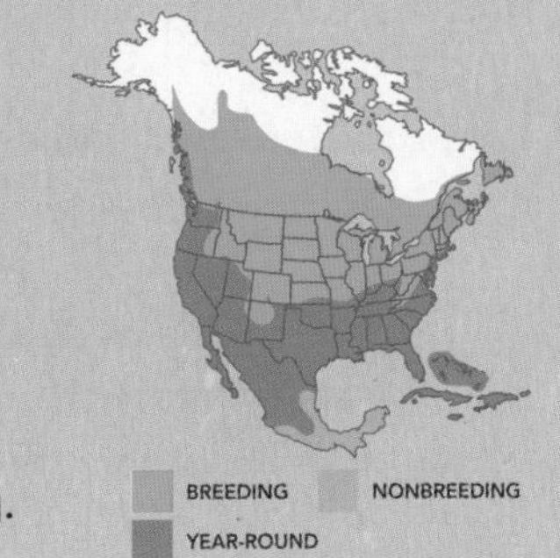

POINT OF FACT It lures predators away from its grounded nest by pretending to have a broken wing.

I SAW IT!

WHEN I SAW IT

DATE

WHERE I SAW IT

SPECIFIC LOCATION, INCLUDE STATE

WHAT IT WAS DOING

NOTES

OPEN COUNTRY BIRDS

Eastern Bluebird

(Sialia sialis)

SIZE 7 inches (18 cm) beak to tail

SHAPE Medium-size, plump perching bird with a large head, long wings, a short tail, and short legs.

COLOR Blue wings, back, and head with a rusty-red throat and chest. Females are dull with more gray than blue and a washed-out, rusty chest.

BEHAVIOR Perches at attention on fences and low branches while softly singing *tur, tur, turley, turley.*

HABITAT Tree-lined meadows, pastures, parks, roadsides, farmland edges.

POINT OF FACT Males attract mates by going in and out of nest cavities with nesting materials in their beaks, all while waving their wings.

I SAW IT!

WHEN I SAW IT

DATE

WHERE I SAW IT

SPECIFIC LOCATION, INCLUDE STATE

WHAT IT WAS DOING

NOTES

Northern Bobwhite

(Colinus virginianus)

SIZE 9 inches (23 cm) beak to tail

SHAPE Small quail with a round body, a small head, and a short tail.

COLOR Body is a busy pattern of brown, rust, cream, and black. Males have black and white masks. Female have muted, tan masks.

BEHAVIOR Flocks run on the ground and burst into quick-flapping flights when alarmed.

HABITAT Open forest, overgrown fields, scrubland.

POINT OF FACT During the spring and summer breeding seasons, males says their names by whistling out *bob-bob-white* as many as two to three times a minute.

I SAW IT!

WHEN I SAW IT
DATE

WHERE I SAW IT
SPECIFIC LOCATION, INCLUDE STATE

WHAT IT WAS DOING

NOTES

....

....

OPEN COUNTRY BIRDS

Black-billed Magpie

(Pica hudsonia)

SIZE 20 inches (51 cm) beak to tail

SHAPE Large perching bird with a very long tail, short wings, and a heavy beak. Its tail makes a diamond shape in flight.

COLOR Black and white with a shiny-blue iridescence on its wings and tail. It has a black beak, black eyes, and black legs.

BEHAVIOR Flashes and forages in busy, noisy flocks while trilling, cackling, and whistling.

HABITAT Open country with brush or scattered trees, farmland, pastures, towns, roadways.

POINT OF FACT Eats fruit, insects, grain, and small vertebrates, and scavenges for dead animals, or carrion, including roadkill.

I SAW IT!

WHEN I SAW IT
DATE

WHERE I SAW IT
SPECIFIC LOCATION, INCLUDE STATE

WHAT IT WAS DOING

NOTES

American Kestrel

(Falco sparverius)

Gray head-band. Black stripes below eye.

SIZE 10 inches (25½ cm) beak to tail

FEMALE

SHAPE Small, slim falcon with a large head, narrow wings, and a long, wide tail.

COLOR Males have gray-blue wings and pale, rusty chests with spots. Females are a streaky brown color with rusty, black-striped backs.

BEHAVIOR Perches on utility wires and fences, and hovers in midair to hunt.

HABITAT Deserts, grasslands, meadows.

POINT OF FACTS Dark, eye-like spots on the back of its head fool predators into thinking it's ready to defend itself.

I SAW IT!

WHEN I SAW IT
DATE

WHERE I SAW IT
SPECIFIC LOCATION, INCLUDE STATE

WHAT IT WAS DOING

NOTES

....

....

OPEN COUNTRY BIRDS

Field Sparrow

(Spizella pusilla)

Rusty crown and splotch behind eye.

SIZE 5½ inches (14 cm) beak to tail

White eye ring.

SHAPE Slender, small sparrow with a round head, a long tail, and a short, cone-shaped beak.

COLOR Brown back with black and white streaks, a gray rump, belly, and chest, and a gray head with a rusty crown and pink beak.

BEHAVIOR Male perches in the open while singing a loud song that speeds up like a dropped ping-pong ball.

HABITAT Tall grassy areas, old farm fields, brushy edges with shrubs and thorny bushes, overgrown meadows.

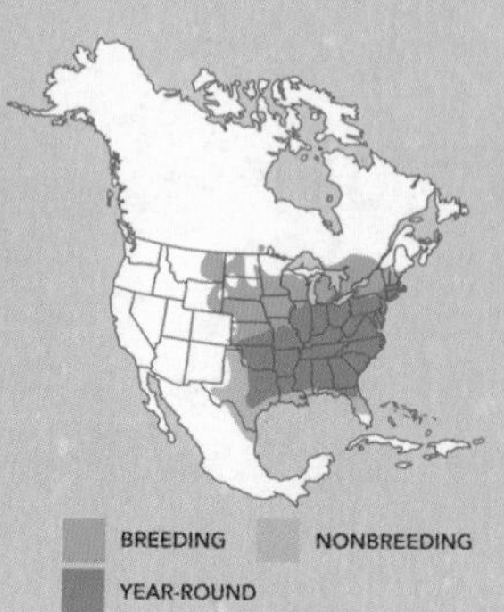

POINT OF FACT Early spring nests are built on the ground, and white-tailed deer sometimes eat the eggs and nestlings.

I SAW IT!

WHEN I SAW IT

DATE

WHERE I SAW IT

SPECIFIC LOCATION, INCLUDE STATE

WHAT IT WAS DOING

NOTES

....

....

White-tailed Ptarmigan

(Lagopus leucura)

Red eyebrow.

Small beak.

No black on tail.

SUMMER

SIZE 11 inches (28 cm) beak to tail

SHAPE Medium, chicken-like bird with a round body, a small head, a short, round tail, and short legs.

WINTER ADULT

COLOR In summer, males are a mottled gray-brown with white bellies. Females are more yellow-brown. Adults are all white in the winter.

BEHAVIOR Forages on the ground to find and eat plants and insects.

HABITAT Rocky mountain tundra, alpine meadows, all above tree line.

POINT OF FACT They are masters of camouflage by blending in with tundra rocks in the summer and snow in the winter.

I SAW IT!

WHEN I SAW IT

DATE

WHERE I SAW IT

SPECIFIC LOCATION, INCLUDE STATE

WHAT IT WAS DOING

NOTES

....

....

Golden Eagle

(Aquila chrysaetos)

SIZE 30 inches (76 cm) beak to tail

SHAPE Huge raptor with a hooked beak, a very wide wingspan with finger-like tips, a small head, and a big tail.

COLOR Dark brown with a golden tint on the back of its head and neck. Juveniles have white patches on their wings and bases of their tails.

BEHAVIOR Soars steadily by creating a shallow V with its wings. It hunts for prey by the air or from a perch.

HABITAT Open country near hills, mountains, rivers, and cliffy areas, tundra, desert, prairies, coniferous forests.

POINT OF FACTS Its natural habitat is above the equator, and it is the national bird of Albania, Germany, Austria, Mexico, and Kazakhstan.

I SAW IT!

WHEN I SAW IT
DATE

WHERE I SAW IT
SPECIFIC LOCATION, INCLUDE STATE

WHAT IT WAS DOING

NOTES

MOUNTAIN BIRDS

Steller's Jay

(Cyanocitta stelleri)

SIZE 12 inches (30½ cm) beak to tail

SHAPE Large perching bird with a large, crested head, rounded wings, a large beak, and a long tail.

COLOR Dark-blue body with a black head, throat, and shoulders. Juveniles are pale with smaller crests.

BEHAVIOR Scolds noisily while flying and foraging in forest treetops.

HABITAT Coniferous forests, parks, backyards.

POINT OF FACT It hops on the ground to investigate sites to stash acorns and pine seeds for the winter. Many of the seeds are never eaten and end up growing into trees!

I SAW IT!

WHEN I SAW IT
DATE

WHERE I SAW IT
SPECIFIC LOCATION, INCLUDE STATE

WHAT IT WAS DOING

NOTES

....

....

MOUNTAIN BIRDS

American Dipper

(Cinclus mexicanus)

SIZE 7 inches (18 cm) beak to tail

SHAPE Medium-size, stocky perching bird with a large head, long legs, and a short tail.

COLOR Dark gray all over, pale gray legs, and white eyelids. Juveniles have yellow beaks and streaky bellies.

BEHAVIOR Flies fast and low over mountain streams. It hops along on rocks to dip its head under the water for aquatic insects.

HABITAT Mountain, coastal, and desert streams with rocky bottoms.

POINT OF FACT This is the only perching bird in North America that's truly aquatic. It swims and catches food underwater.

I SAW IT!

WHEN I SAW IT

DATE

WHERE I SAW IT

SPECIFIC LOCATION, INCLUDE STATE

WHAT IT WAS DOING

NOTES

Mountain Bluebird

(Sialia currucoides)

SIZE 7 inches (18 cm) beak to tail

SHAPE Medium-size perching bird with a large head, a thin beak, a long tail, and long wings.

COLOR Pale blue overall with white under its tail. Females are gray-brown with blue tints on their wings and tails.

BEHAVIOR Hovers in flight while focusing on tasty ground insects to hunt.

HABITAT Open country with scattered trees or sagebrush, mountain meadows, prairies, sagebrush, alpine tundra.

POINT OF FACT Choosing a mate is all about the cavity nest the male can offer. Females don't care about his singing or his good looks.

I SAW IT!

WHEN I SAW IT

DATE

WHERE I SAW IT

SPECIFIC LOCATION, INCLUDE STATE

WHAT IT WAS DOING

NOTES

Common Raven

(Corvus corax)

Shaggy throat feathers.

Huge beak.

SIZE 23 inches (58½ cm) beak to tail

SHAPE Huge perching bird with a thick neck, a massive beak, and a long, wide tail. It has thin, finger-shaped wingtips during flight.

COLOR Black all over, including legs, eyes, and beak.

BEHAVIOR Flies alone or in pairs with slow flaps and smooth glides. It struts and hops with both of its feet on the ground to investigate.

HABITAT Open high country, coniferous forests, tundra, seacoast, towns.

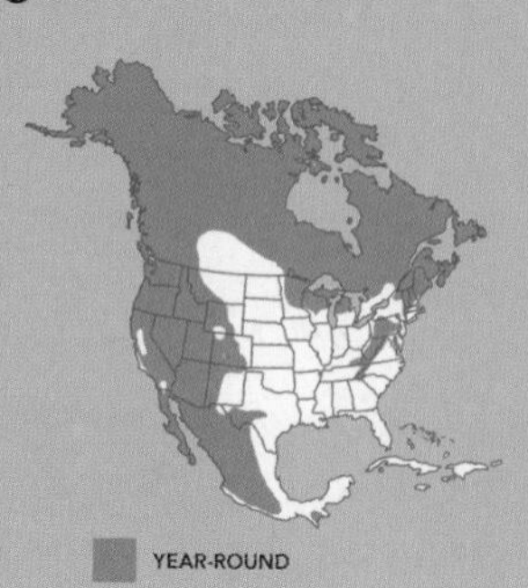

POINT OF FACT Intelligent and curious, it is able to learn faces, open locks, and work with others in a team.

I SAW IT!

WHEN I SAW IT

DATE

WHERE I SAW IT

SPECIFIC LOCATION, INCLUDE STATE

WHAT IT WAS DOING

NOTES

Western Tanager

(Piranga ludoviciana)

SIZE 7 inches (18 cm) beak to tail

SHAPE Medium, full-bodied perching bird with a short beak and a medium-length tail.

COLOR Yellow with black wings and an orange-red head and throat. Females and juveniles are a dull yellow overall with dark wings.

BEHAVIOR Forages among leaves and coniferous needles on high tree branches.

HABITAT Coniferous forests during breeding season; woodlands and forest edges.

POINT OF FACT Sings a hoarse, robin-like *brr-eet, burry, burry, brr-eet.*

I SAW IT!

WHEN I SAW IT

DATE

WHERE I SAW IT

SPECIFIC LOCATION, INCLUDE STATE

WHAT IT WAS DOING

NOTES

....

....

Dusky Grouse

(Dendragapus obscurus)

SIZE 18 inches (45½ cm) beak to tail

SHAPE Large, chicken-like bird with a small head and a long tail.

COLOR Dark, blue-brown with yellow eyebrows and an orange, bare neck patch. Females are a dark, brownish-gray with white flecks.

BEHAVIOR Forages on the ground for leaves and in coniferous trees for needles.

HABITAT Coniferous forests in winter; dry grasslands and scrubland in summer.

POINT OF FACT When threatened, it often freezes in place to avoid being seen. Its camouflage coloring helps it go unnoticed.

I SAW IT!

WHEN I SAW IT

DATE

WHERE I SAW IT

SPECIFIC LOCATION, INCLUDE STATE

WHAT IT WAS DOING

NOTES

....

....

MOUNTAIN BIRDS

Clark's Nutcracker

(Nucifraga columbiana)

SIZE 11 inches (28 cm) beak to tail

SHAPE Fairly large perching bird with a rounded head, a short tail, and a long, sharp beak.

COLOR Light gray with black beak, wings, eyes, and legs. In flight, it appears to have white sides on its tail and wing ends.

BEHAVIOR Flocks gather in pine trees to pry seeds from cones. It flies with swooping flaps by flashing its wings and tail.

HABITAT Mountain coniferous, especially pine forests.

POINT OF FACT Seeds that it pries from pinecones are held in a pouch under its tongue to be carried away and buried for the winter.

I SAW IT!

WHEN I SAW IT

DATE

WHERE I SAW IT

SPECIFIC LOCATION, INCLUDE STATE

WHAT IT WAS DOING

NOTES

....

....

American Three-toed Woodpecker

(Picoides dorsalis)

SIZE 8 inches (20½ cm) beak to tail

SHAPE Medium woodpecker with a long beak and three toes.

COLOR Black-and-white body, a black head with a white mustache stripe, a thin, white eyeline, and a yellow patch on its crown. Females don't have crown patches. Juveniles have small, yellow crown patches.

BEHAVIOR Strips the bark on tree trunks to forage for insects, or looks for them on fallen logs.

HABITAT Mountain and alpine coniferous forests.

POINT OF FACT Unlike most woodpeckers, they only have three toes!

I SAW IT!

WHEN I SAW IT
DATE

WHERE I SAW IT
SPECIFIC LOCATION, INCLUDE STATE

WHAT IT WAS DOING

NOTES

....

....

MOUNTAIN BIRDS

Pygmy Nuthatch

(Sitta pygmaea)

SIZE 4 inches (10 cm) beak to tail

SHAPE Tiny perching bird with a large head, a straight beak, and a short, square tail.

COLOR Gray wings and back, a brown cap, and a cream to tan throat, chest, and belly.

BEHAVIOR Acrobatically climbs pine tree trunks and branches, and clings upside-down from cones.

HABITAT Western pine forests.

POINT OF FACT Constantly calls a squeaky, high-pitched *pee-deedee, pee-pee, pee-deedee* that sounds like a squeezed rubber ducky or dog toy.

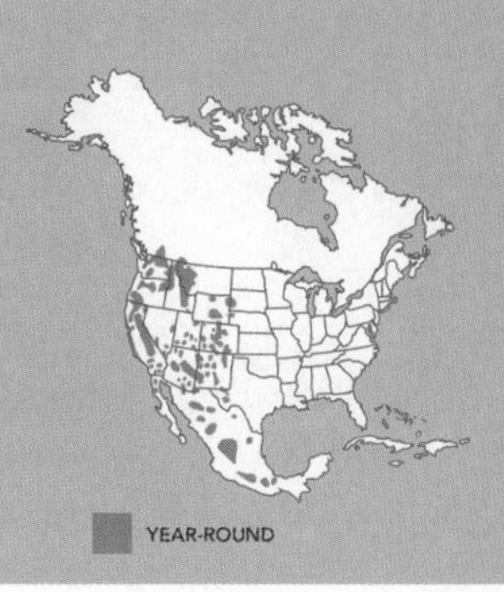

I SAW IT!

WHEN I SAW IT

DATE

WHERE I SAW IT

SPECIFIC LOCATION, INCLUDE STATE

WHAT IT WAS DOING

NOTES

....

....

Gray-crowned Rosy Finch

(Leucosticte tephrocotis)

SIZE 7 inches (18 cm) beak to tail

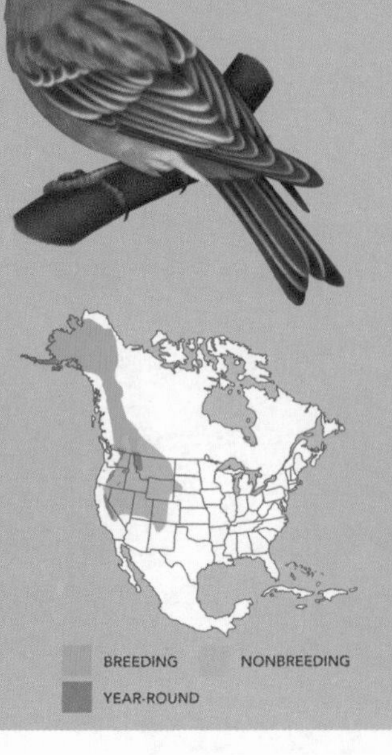

SHAPE Small perching bird (but medium for a finch!) with a round shape, a thick neck, and a cone-shaped beak.

COLOR Dark brown overall with a gray head and a pink tint on its belly, rump, and wings. Juveniles are brown all over.

BEHAVIOR Fearlessly forages for insects and seeds on the ground, in snowfields, and in conifers.

HABITAT Western mountain meadows and extreme alpine areas.

POINT OF FACT Few of their nests have been seen by humans because they breed in remote places within very harsh environments.

I SAW IT!

WHEN I SAW IT
DATE

WHERE I SAW IT
SPECIFIC LOCATION, INCLUDE STATE

WHAT IT WAS DOING

NOTES

....

....

Mountain Chickadee

(Poecile gambeli)

SIZE 5 inches (12½ cm) beak to tail

SHAPE Very small perching bird with a long, narrow tail and a small beak.

COLOR Gray back and wings, cream undersides, and a black-and-white head.

BEHAVIOR Busily flits in small flocks among high branches, clinging to twigs and pinecones calling *chickadee-dee-dee*.

HABITAT Western mountain coniferous and deciduous forests.

POINT OF FACT A female on a nest will lunge forward and hiss like a snake to scare off predators.

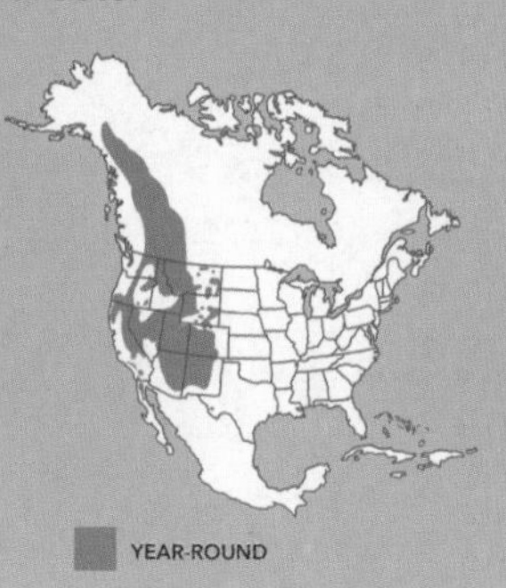

I SAW IT!

WHEN I SAW IT
DATE

WHERE I SAW IT
SPECIFIC LOCATION, INCLUDE STATE

WHAT IT WAS DOING

NOTES

....

....

Golden-crowned Kinglet

(Regulus satrapa)

SIZE 3¾ inches (9½ cm) beak to tail

SHAPE Tiny, round-bodied perching bird with short wings and a thin tail.

COLOR Green back, a gray chest and belly, a black-and-white face, and an orange crown patch. Females have yellow crown patches.

BEHAVIOR Keeps hidden high up in trees, singing a high-pitched *seet-seet-seet*.

HABITAT High mountain forests in summer and mixed forests, swamps, towns in winter.

POINT OF FACT A single miniature feather covers each of its nostrils.

BREEDING NONBREEDING
MIGRATION YEAR-ROUND

I SAW IT!

WHEN I SAW IT
DATE

WHERE I SAW IT
SPECIFIC LOCATION, INCLUDE STATE

WHAT IT WAS DOING

NOTES

....

....

White-crowned Sparrow

(Zonotrichia leucophrys)

SIZE 6¼ inches (16 cm) beak to tail

SHAPE Small, long perching bird (but large for a sparrow!) with a flat head, a small beak, and a long tail.

COLOR Gray chest, belly, and throat, brown wings, and a black-and-white head.

BEHAVIOR Small flocks hop on ground, in bushes, or low branches looking for seeds.

HABITAT Scrubby, brushy areas near fields, forest edges, thickets.

POINT OF FACT Learns its song as a youngster from other birds around it, so their songs differ by region, almost like accents!

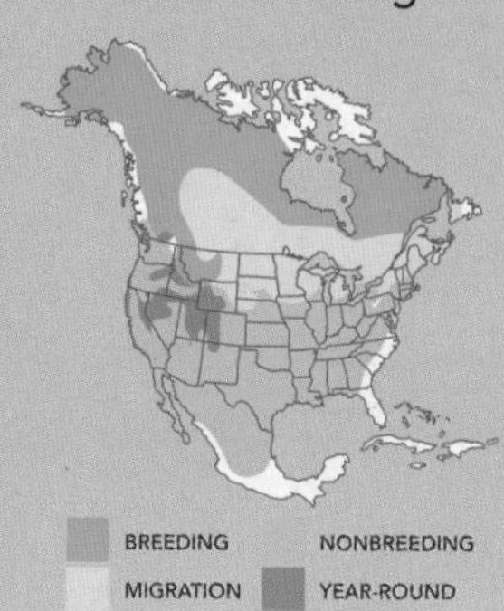

I SAW IT!

WHEN I SAW IT

DATE

WHERE I SAW IT

SPECIFIC LOCATION, INCLUDE STATE

WHAT IT WAS DOING

NOTES

....

....

PART III

MAMMALS

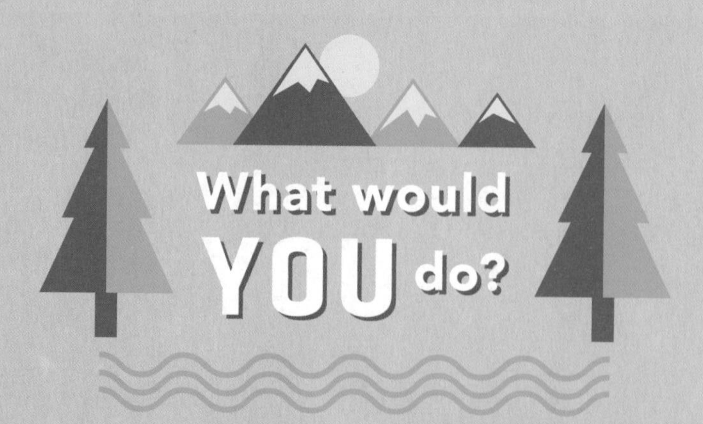

You've been hiking all morning, so you stop for a snack. There's nothing like the green-on-green of a high-country river valley. You settled onto a soft mossy spot to eat, when all of a sudden you're attacked—by a nap!

You wake to the sound of crunching plants. It's a shiny brown moose calf! The youngster stands in the marshy grass just fifteen feet away. You quietly sit up and reach for your camera when a loud, breathy, sound comes out of the trees behind you.

Hurmph-hurmph!

Hurmph-hurmph!

It's mamma moose. And you're between her and the calf! Should you lie back down and pretend to be asleep? Or will you get trampled? Should you get up and run? Or will that get you kicked? *What would you do?*

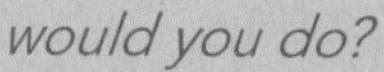

CHAPTER 1

How to Spot Mammals

Witnessing a bear munching berries or a fox on patrol is awe-inspiring. It feels like glimpsing a secret world. Another reason spotting wild furry animals is exciting? It makes you feel lucky! Because, while wild animals might live nearby, seeing them isn't always easy. So how do you find and spot wild animals? The first tip is knowing your target. The more you know about the animals you want to see, the better the chance you'll succeed. Let's get started.

BLACK BEARS

What Makes a Mammal?

Animals with fur are mammals. And, yes, that includes you. Mammals are warm-blooded vertebrate animals, as are birds. But instead of feathers, mammals are covered in fur or hair. (Though sometimes it's just a few whiskers, as with whales and naked mole rats.) A mammal mother also gives birth to young that drink milk her own body makes.

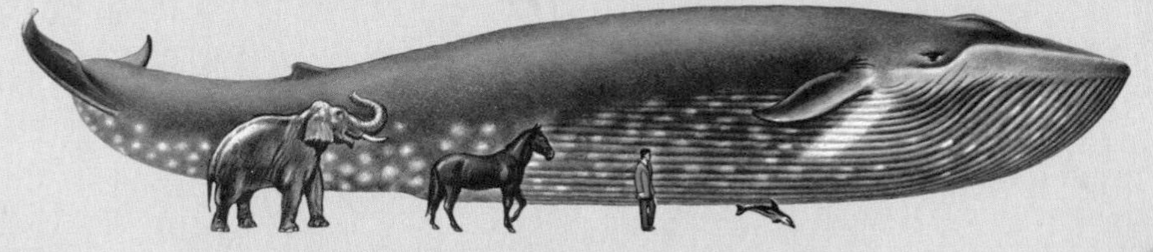

Mammals come in a wide range of sizes and shapes.

Young that are nursed get more care and protection from parents than newly hatched snakes or fish babies. What else is special about furry animals? The brains of mammals are the largest and most developed among animals. An elephant's brain is four times the size of the one you're carrying around.

Earth's mammals have been evolving for more than 200 million years.

About 450 different species of wild mammals make North America their home. Everything from bison and elk, pumas and seals, to moles and squirrels. Most have four legs, a mouth full of teeth, and a coat of fur. But a gray whale has none of these! And seals have flippers, not feet, while bats have wings for arms. Magnificent mammals have evolved a variety of adaptations for surviving in different environments—whether they're prairie grazers, desert burrowers, mountain stalkers, forest gatherers, or ocean hunters.

Mammal Scouting Tips

Why are mammals harder to see than birds? Many mammals aren't active when people are. They're nocturnal, or at least crepuscular (a fancy word for being active at dawn and dusk). Most mammals try *not* to be seen, hiding in holes and trees, under bushes and behind rocks. Hiding is how tasty prey animals, like mice, avoid becoming someone's dinner. And staying unseen makes for smart stalking by hunting mammals, like foxes.

So how can you up your odds of seeing wild mammals? Spotting strategies depend on who you're looking for, of course. Seeing a sea lion takes a trip to the coast, and bat watching is a nighttime hobby. But here are some mammal-specific wildlife-watching tips to add to those on page 194 when looking for land-living, four-legged mammals.

Wearing camo, keeping a low profile, and staying downwind all help when you're out wildlife watching for mammals.

GET UP EARLY Dawn is when many mammals are foraging and/or finishing up their nighttime activities.

TRY TO BLEND IN If you've got camo, wear it. Or dress in drab, earthy colors that match wherever you're going.

GO LOW PROFILE There's a lot less of you to notice if you're crouched behind a rock or bush than if you're standing out in the open. Keep low, and use the landscape to help you hide.

EARS OPEN, MOUTH CLOSED Most mammals hear (and smell) better than they see, especially at a distance. Try to walk and move quietly, too. It'll help you listen better.

THINK ABOUT WIND Mammals are super smellers. Being positioned so the wind is blowing from the animal toward you—downwind—can make you invisible. Test it: Toss a bit of dust into the air. Does it blow toward you, or away from you? If it blows in the direction you're looking, then whoever's there can smell you!

SKIP THE SCENTED STUFF Body sprays and perfumes are red alerts that humans are around.

Seek the Signs

Fortunately you don't have to actually see a deer, bat, chipmunk, or other mammal to know it's around. They leave clues for you, called signs. And mammal signs are everything from scat to skulls, hair clumps to holes, chewed nuts to clipped plants. Wildlife watchers pay attention to signs to find out who's around—and up their chances of spotting them. Mammal signs are everywhere—on the ground, up in trees, under rock ledges. Train your eyes to be on the lookout not just for mammals, but mammal signs like these:

TRACKS Paw prints, hoofprints, and tail drags, too.

SCAT Everybody poops! But it doesn't all look the same.

TRAILS AND RUNWAYS Everyone from foraging mice to traveling deer create paths as they travel routine routes.

BURROWS, HOLES, AND DENS Whether chewed out, dug out, or piled up, someone likely lives there.

FUR Just like in the bathtub, hair gets left around.

LEFTOVERS Nutshells, stripped pinecones, nibbled buds, chewed bark, clipped plants, or scraps of rabbit pelt mean someone's been snacking.

SOUNDS Squirrels churr, chipmunks chirp, deer huff, coyotes yip, bobcats scream—hearing is believing, too.

SMELLS Skunk scent, stinky bat guano, and urine odors are all mammal clues, too.

BONES AND BITS Skulls, teeth, bones, horns, claws, and shed antlers all came from someone.

CLAW, ANTLER, AND TOOTH MARKS Bears mark territory with claws, chewed-up stumps are signs of beaver, and antlered animals leave small trees with ribbons of bark.

CAUTION!

Wild mammals can be dangerous! They've got biting teeth, scratching claws, and hooves that kick. They can also carry diseases. Raccoon, fox, and bat bites can transmit rabies. (Wildlife workers are vaccinated.) Rodents spread hantavirus and plague. Don't get too close, and never touch a wild animal. Read more about how to safely spot animals on page 194.

TRY IT → Seek the Signs

Don't see any furry wildlife running around? Look closer. There are likely clues that some mammal has been here.

WHAT YOU'LL NEED

- a pencil or pen

STEP 1 Take a walk or hike in a park or other nature spot.

STEP 2 Check off the mammal signs you encounter.

Sights:

☐ tracks in mud or sand ☐ scat or droppings

☐ hole or burrow ☐ chewed-up acorns or nuts

Anything else? ....

Sounds:

☐ churr, chip, or chirp of squirrel or chipmunk

Anything else? ....

Smells:

☐ skunky, musky ☐ guano-ish ☐ pee stink

Anything else? ....

I DID IT! DATE:

TRACK IT ↘ Record the Sign

Make a record of the mammal sign you saw by drawing and tracking it below.

DATE

TIME

LOCATION

WEATHER

STEP 1 What did you see?

STEP 2 Draw it.

STEP 3 Any guesses on who made the sign?

STEP 4 See any actual mammals while drawing or writing? Did you recognize it or know its name? Look for it (or them!) in the Mammal Identification section (pages 240–301). Make a positive ID? Check off its I SAW IT! box and fill in the blanks. You're a wildlife watcher now!

I DID IT! DATE:

TRACK IT ↘ Mapping Sound

Listening can be learned. Like any other skill, it just takes practice. Make a sound map with a friend to compare skills.

WHAT YOU'LL NEED

> a crayon or marker, two sheet of paper, a friend or family member, and a watch, clock, or timer

STEP 1 Find a quiet place outside.

STEP 2 Write your initials in the center of the paper. Ask your friend to initial the center of his or her paper as well.

STEP 3 Sit on the floor (or ground) back-to-back with your friend. Set the initialed paper on the ground (or floor) in front of you. Your initials represents you, the listener, where you're sitting.

STEP 4 When you hear a sound, draw a small X on the page relative to you in distance and direction. For example: If the sound is very far away and behind you, then put the X on the page below your initials at the edge of the blank space. If it's loud, make a thick X. A faint sound? Draw a light X. Got it?

STEP 5 Listen quietly for ten minutes, marking down Xs for the sounds you hear.

STEP 6 Compare sound maps to find out who heard what. Did the two of you hear the same sounds at the same volumes?

I DID IT! DATE:

TRACK IT ↘

EASTERN GRAY SQUIRREL

It Sounded Like What?

Sounds can be difficult to describe. But writing down what you hear can help you remember it—and share it with others. The next time a squirrel chatters or raccoons noisily root around the garbage cans, listen carefully. Then track the sound below.

DATE

TIME

LOCATION

WEATHER

1. What did you hear? Describe it:

2. Did you see what was making the sound?

3. Any guesses on who you heard or what it was?

4. Did you actually *see* the sound's maker? Did you recognize it or know its name? Look for it in the Mammal Identification section (pages 240–301). Make a positive ID? Way to go! Check off its I SAW IT! box and fill in the blanks. Write what you heard under notes.

I DID IT! DATE:

TAKE IT TO THE NEXT LEVEL ↗

Sound Scavenger Hunt

Listen up when you're outside for mammal-made sounds! Write in the mammal you saw making these sounds:

whistle:

chipping:

snort:

chirring:

Spot the mammal, but can't identify it? No worries! Use the Mammal Identification section (pages 240–301) for help. Check under *Behavior* for "sound advice."

CHAPTER 2

How to ID Furry Friends

Now that you've learned some strategies for spotting North American mammals, you'll want to identify those you see. Don't worry! You already know a bunch of them. There's no mistaking a bison, for example. Same goes for a raccoon, armadillo, and opossum. They're not like anything else on this continent, so there's not much chance for confusion. Whew!

RACCOON

Other mammals are trickier to ID. There are dozens of kinds of chipmunks, twice again as many shrews, and forty plus species of bats.

NINE-BANDED ARMADILLO

So how can you tell them apart? Like with any kind of wildlife you can use these five clues to identify mammals.

Mammal Identification 101

SHAPES OF MAMMALS While there may be dozens of species of bats, you know a bat when you see one. Its characteristic wings give it a unique shape. Same goes for many of the major mammal groups. They're familiar animals—rabbits and rats, squirrels and bears, deer and seals—with recognizable shapes and giveaway traits. And even the groups you might be less familiar with are easily distinguished once you know the one or two telltale (oops, tell-*tail*) characteristics that give the group away. Difference between ground squirrels and tree squirrels? Tree squirrels have a bushy tail. Seals from sea lions? Sea lions have ear flaps. Shrews from mice? Shrews have tiny eyes and ears.

A tree squirrel's long, bushy tail (above) is used to balance on branches—something a ground squirrel (below) doesn't do as much.

SIZE Knowing general sizes helps. A large mouse still runs from a small fox. But if you're trying to identify a type of mouse, knowing which ones are large or small is important. Being able to say an animal is 5 or 15 inches (12½ or 38 centimeters) in length can be tough. Relative sizes work, too. Is it about as long as a soda can? The size of a concrete block? Or measure yourself! Is the mouse larger than your fist? Would the deer's shoulder be higher than yours? It's a lot easier to measure your foot or a cinderblock than a woodchuck in the woods!

5 INCHES (12½ CM)

16 INCHES (40½ CM)

28 INCHES (71 CM)

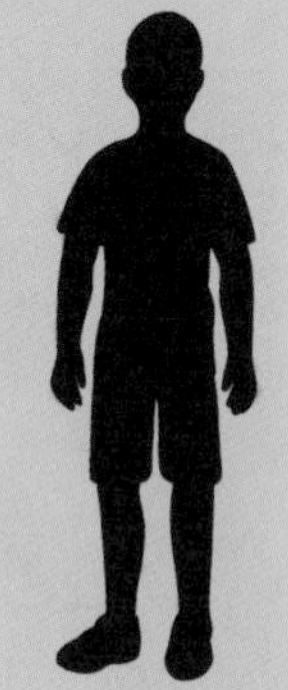

58 INCHES (147½ CM)

COLOR & COLOR PATTERNS Mammals aren't the most colorful wildlife (there's lots of brown), but color is still important to identification. Especially when it comes to stripes, rings, rump patches, masks, and other patterns that serve as field marks.

Can you count the stripes on this ground squirrel?

BEHAVIOR A rabbit escaping into the bushes hops, while a woodchuck doing the same scurries away. Nothing screams like a shrew. Moose don't gather in huge herds. And even a quick glimpse of a surfacing dolphin tells you what it is. How an animal moves, sounds, and behaves are big clues to who it is.

BRUSH RABBIT

RANGE & HABITAT If you just saw something fly by while mountain climbing in Montana, it was not a Mexican free-tailed bat. It just wasn't. Same goes for seeing a sea lion in Maine or a kangaroo rat in a swamp—it'll simply never happen. This is why field guides have range maps. You can often eliminate possibilities by checking who lives where you are. Sometimes it's the most important element of identification. If you're ever lucky enough to see a flying squirrel, you'll more easily identify the species (southern or northern) based on your location. And congratulations!

Pictures Can Help

Identifying an unknown mammal in real time isn't always possible. That's where pictures come in. After all, wildlife photography is a hobby shared by millions around the world. And photos can help identify animals. But just because you can snap a pic with your phone doesn't mean sketching is obsolete. Why not? Drawing animals trains the eye to observe. Sketching the tail of a chipmunk makes you really look at it and be more likely to remember what it looks like.

YELLOW-PINE CHIPMUNKS

Recognizing Eyeshine

Because so many mammals are out and about after dak, eyeshine can be a helpful clue for wildlife spotters. Eyeshine is light reflecting off special tissue at the back of some animals' eyes. You've probably seen animal eyes glowing in the dark before. The mirrorlike layer is called the tapetum and occurs in the eyes of mostly nocturnal and crepuscular animals. (Nope, not your eyes.) Having reflecting eyes is an adaptation for seeing in dim light. The tapetum bounces light entering the eye back past the light receptors a second time, doubling the amount of light captured. That's super useful to a puma hunting after dark and a rabbit trying to eat a late night snack without becoming one.

The color, size, and shape of eyeshine differs among animal species. It's not always an accurate field mark because eyeshine color can vary among individuals depending on the age and health of the animal, as well as the kind of flashlight you're using. But eyeshine is one clue in identifying an animal—and it's definitely a mammal sign.

Talk Like a Mammalogist

Here are some terms to know when swapping stories with other mammal spotters.

CANINES OR CANIDS dogs, including foxes, wolves, and coyotes

CETACEANS whales, dolphins, and porpoises

GRAY WHALE

CREPUSCULAR active at dawn and dusk

DIURNAL active during day

FELINES OR FELIDS cats, including pumas and bobcats

FIELD MARK distinctive feature useful in identification

HAULOUTS places along shores where seals and sea lions gather and rest

LAGOMORPHS rabbits, hares, and their non-rodent kin

MARSUPIALS mammals that carry young in pouches, like kangaroos

AMERICAN MINK

MUSTELIDS weasels, badgers, and their kin

NOCTURNAL active at night

PINNIPEDS seals, sea lions, walruses, and their relatives

RODENTS gnawing mammals, including mice, rats, squirrels, and their relatives

SEMIAQUATIC partially living in water

SOCIAL lives in groups

SOLITARY lives alone except for breeding

EASTERN FOX SQUIRREL

UNGULATES deer, moose, sheep, and other hoofed mammals

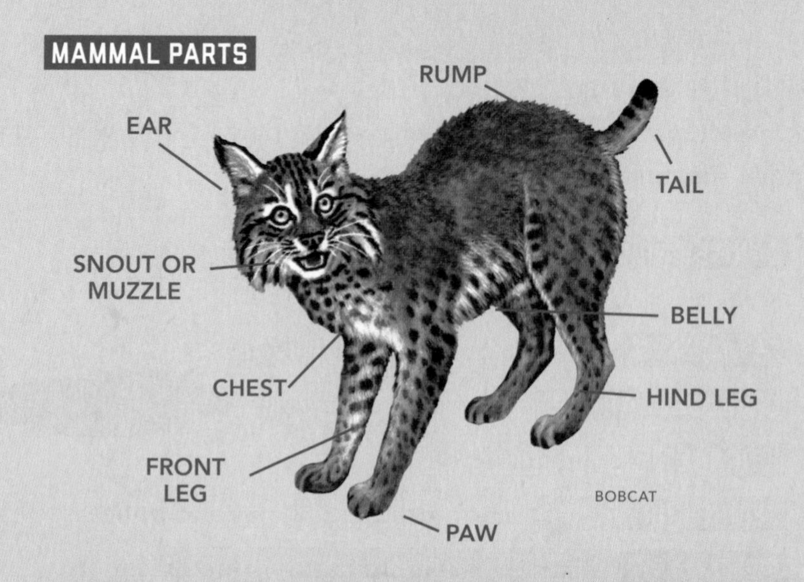

CRITTER CONFUSION: Grizzly or Black Bear?

See how the grizzly has a higher shoulder and bigger head? Shape is more reliable than color when it comes to distinguishing bears. Being able to recognize the difference in shape between a black bear and grizzly bear is *bear-y* important. Why? How you should react to a bear encounter depends on the kind of bear you run into. Black bears can be scared off with noise, standing tall, yelling, and looking big. But that's the wrong strategy with a grizzly. It is more likely to see that sort of behavior as a threat that calls for aggression in return. Whatever the bear, NEVER run. That makes you chase-worthy prey.

TRY IT → Spot a Mammal

Squirrels are a great starter mammal to search for—they are probably the easiest wild mammal to find and observe. Tree squirrels and ground squirrels (including chipmunks) are common, diurnal, and often coexist with people in neighborhoods and parks. Lucky enough to live near a sea lion haulout, or where you regularly see deer? That'll work, too.

CALIFORNIA SEA LION

WHAT YOU'LL NEED

➢ pencil and/or pen, binocs, camera (optional)

STEP 1 Find a wild mammal to watch and settle in.

STEP 2 Record and track information about it below as you watch.

STEP 3 Can you identify the mammal from your notes and observations? Use the Mammal Identification section (pages 240–301) and/or a field guide to help. Make a positive ID? Fabulous work! Check off its I SAW IT! box and fill in the blanks.

STEP 4 Once you've got an ID, compare your size estimates, field marks, and other observations against its page in the Mammal Identification section (pages 240–301 or field guide. Any big differences?

DATE

TIME

LOCATION

WEATHER

Group (rabbit, squirrel, seal, etc.)

Estimated length of body and tail (shoulder height if hoofed animal)

Color:

Color patterns or field marks, such as stripes, spots, rings, or anything else?

What's it doing?

How does it move?

Sounds?

Solitary or in a group?

Describe habitat:

I DID IT! DATE:

TRACK IT ↘ Draw a Furry Friend

Sketching animals is a terrific way to train the eye to notice details, which hones wildlife observation skills and boosts memory. The general shape of four-legged furry animals (sorry, whales and bats) can be drawn with three ovals or circles. Sharpen a pencil and give it a try yourself.

WHAT YOU'LL NEED

> white paper, a pencil or pen, a camera (optional)

STEP 1 Notice how reducing the shape of the animal to three circles makes other characteristics pop out? Like how long the deer's legs and neck are, and how big the raccoon's rump is?

STEP 2 You can practice drawing the three circles yourself. Choose an illustration from the Mammal Identification section (pages 240–301). Place a sheet of white paper (the thinner the better!) over the illustration. You need to see the animal's shape through it.

STEP 3 Draw an oval or circle over its rump. Then its shoulder. Now a circle for the head.

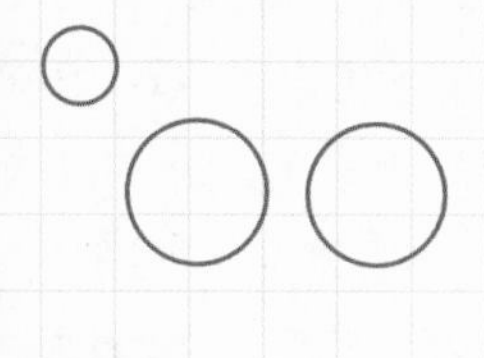

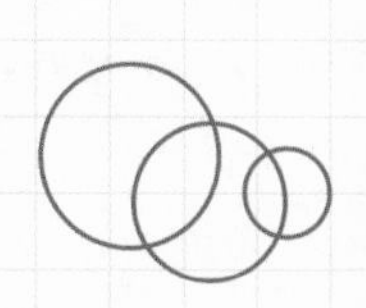

STEP 4 Add in lines for the back, belly, and neck to create an overall outline.

STEP 5 Then sketch in a snout, ears, legs, tail, and other details to finish the drawing. You're a mammal artist!

STEP 6 Now take your skills outside. Find a wild mammal to sketch, or take a photo of one. (It holds still easier that way.) Then repeat steps 3–5, using the three-circle method of mammal drawing to get you started. You're a wildlife artist now!

I DID IT! DATE:

TAKE IT TO THE NEXT LEVEL ↗

The Eyes Have It

Practice spotting eyeshine to identify mammals.

WHAT YOU'LL NEED

- a headlamp with low light setting, a second flashlight, a pencil or pen, colored pencils or markers

STEP 1 Prepare for a night hike. When heading out on the path or trail, wear the headlamp so that the lamp shines from your forehead with the beam pointing straight out. Set it on the lowest setting that still allows you to see well enough to safely walk. Have the other flashlight in your hand at the ready but not turned on.

STEP 2 Quietly walk and scan all around. Some tips:

- Look right and left, toward both sides of the path as well as up and down, at ground level, up in the trees, etc. Not just right in front of you.
- Keep moving; animals aren't going to come to you. Surprise is important!
- Listen! If you hear leaves rustling or something squeak, look in that direction.

STEP 3 When you see a flash of eyeshine, stop, and turn on and point the brighter handheld flashlight toward the eyeshine. Look for:

- **Color:** Yellow-green, orange-red, white?
- **Pupil:** Visible or not? Same color or different?
- **Shape and size:** Almond-shaped? Round? Oval? Pea-size? Big as nickels?
- **Setting:** Far apart like deer's eyes or close together like a raccoon's?

STEP 4 Now think about what animal it might be. Does the guess make sense? You're probably not seeing a deer up in a tree or a coyote disappear down a tiny hole.

STEP 5 Drawing (and coloring) the shapes of the eyeshine you spotted is a good way to memorize this awesome experience.

CHAPTER 3

Mammal Tracking

Want to know which mammals are around—even when you don't see any? Tracks can lead you to mammals you're looking for, tell you who was here recently, and teach you which mammals regularly visit places. Many mammal species can be determined solely (ha!) from the paw prints left behind. Why? The bodies of mammals reflect how they live. Limbs are adapted to flying (wings), running (hooves), swimming (flippers), and digging (claws).

BAT WING

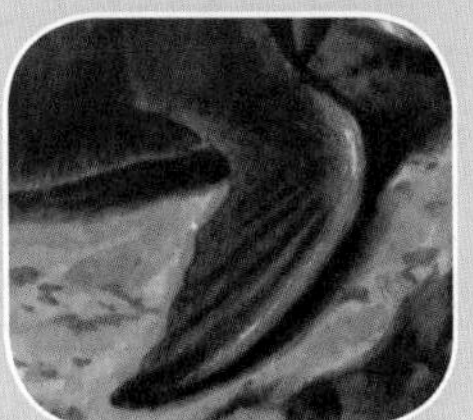

SEAL FLIPPERS

MOLE FRONT FOOT

ELK HOOVES

What Tracks Tell Us

An animal's footprint tells you what the bottom of its foot looks like. And that unique foot shape reflects how the mammal gets around. For example, hopping and leaping mammals like rabbits and squirrels leave tracks with the larger hind feet in front of the front paws. They look a bit like double exclamation points!!

COTTONTAIL

The pattern of paw and hoof pads, claw arrangement, and overall shape and size offer strong clues you can use to identify the track maker.

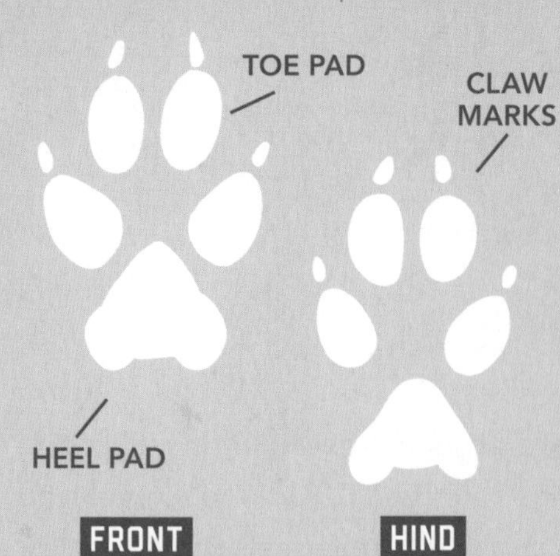

Hooved animals like elk and deer have nearly identical front and back tracks, for example. The claws of coyotes and bears show up on tracks, but not those of bobcats and pumas. Like your house kitty, their claws retract to stay super sharp.

The tracks of some animals don't perfectly match the shape of their feet. Animals with especially hairy feet or that often run on their tiptoes can leave incomplete tracks. Track identification guides, including this book, show the part of the foot that doesn't leave a track as a simple outline.

Stories in Prints

Who was there isn't the only question answered by tracks. The direction, spacing, and depth of tracks tell a story about where the animal was going and how fast. Look around. See another pair of tracks? Who might have been chasing it, or was it alone? Did the animal stop at one point—look around for danger or prey, have a snack, or just catch its breath—and then start walking again? There's a lot to learn from mammal tracks—if you know how to read them.

Track Length and Width

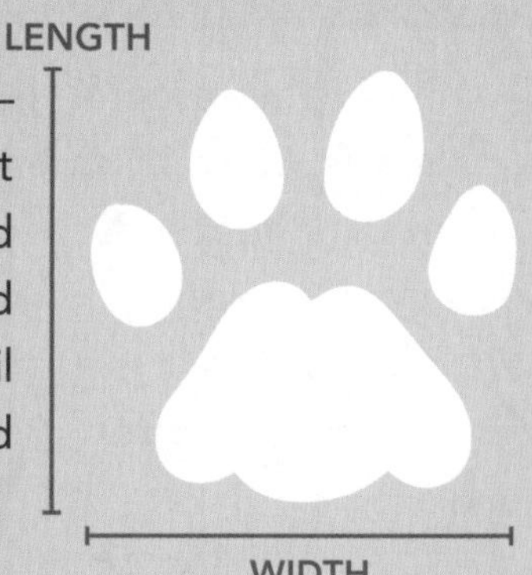

Tracks show up best on wet, firm soils—beaches, along rivers and creeks, moist trails—and in new snow. The dust or sand that covers sunbaked paths in deserts and arid areas can be good for tracks, too. (Until it gets windy!) But you don't need to find

perfect tracks to figure out who made them. One single clear track can be enough to identify an animal. Track length is the size of the paw print from top to bottom. Track width is the size of the paw print from left to right. And a single set of four paws in a row clues you in to what the animal was doing—standing still, hopping, walking, or running. How? The hoof tracks of a running deer look different from those of a walking deer. The two crescents spread apart, splayed by the force of digging into the dirt, shows that the deer was running. Tiny toes called dewclaws on the back of the deer's foot also show up as dots on running tracks.

Straddle and Stride

Imagine a winter morning after a light snowfall. There are tracks everywhere in the meadow! Think about all the patterns you see. Can you tell what way the animals were moving? That's called the line of travel. Every critter makes not only a unique shape of paw print, but creates a characteristic pattern based on its body size, limbs, and mode of movement. A bear cub's tracks looks different from its mom's not only in paw size. Its stride and straddle are smaller, too. Track reading has its own lingo! Stride is the distance between two steps made by the same foot. Straddle is the overall width of a set of tracks.

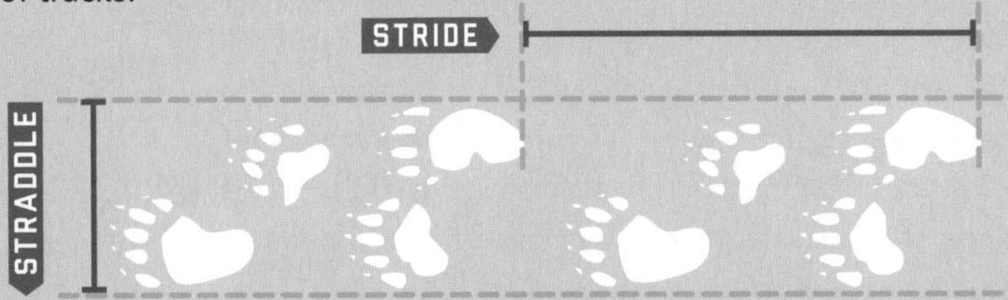

Tracks of All Sorts

Feet aren't the only body part that leaves tracks. Depending on the soil, the tails of some mammals leave marks between the paw prints. These tail drags are another clue to the tracks' maker. Rodents and semiaquatic mammals like otters and beavers are famous for leaving tail drags between tracks.

Now that you know a bit about mammal tracks and how they vary, flip through the pages of the Mammal Identification section (pages 240–301). Each of the four-legged mammal pages features an illustration of the animal's tracks. Do the tracks match the animal as you thought they would? Below are three of the most surprising. Do you know whose tracks they are?

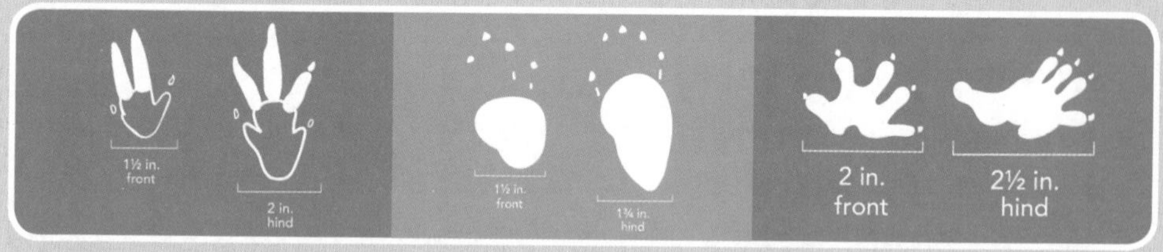

From Tracks to Trails

Many animals travel daily along the same pathways between den, foraging or hunting grounds, and a source of water. Over time all those hooves and paws wear a path through the grass, make a trail in the woods, or create a tunnel under bushes. You've probably seen some of these game trails, animal runways, or wildlife traces while out hiking. Deer trails can be wide enough to fool a hiker into leaving the real trail! Beavers and muskrats create trails that disappear into ponds and streams. Runways made by mice and other small rodents look like tunnels through grass or paths along sheds and fences. Moles leave a raised ridge or dirt as they tunnel under the surface.

RIVER OTTER

TRY IT → Check Out Animal Tracks

Get familiar with the different kinds of mammal tracks out there.

STEP 1 Go for a walk or hike. Try to wander someplace that has different kinds of habitat, like a trail in the wood along a stream, or a park with trees and woods.

STEP 2 Look where the soil is sandy, muddy, or dusty for tracks. You're in luck if there's new snow!

STEP 3 When you see tracks, take a closer look. Here are some things to look for:

→ What is the shape and size?

→ Are there toes? Count them. (Bears have five, in case you're wondering.)

→ Can you see marks from claws poking into soil?

→ How different are the front and hind paws?

→ Anything unusual about the tracks? Webbed feet (beaver, otter), thumb-like toe (opossum), etc.

STEP 4 Take another look at the entire track. How would you describe it to someone else? For example: *a wide track with round heel pad, five toes, and claw marks.* This trains your brain to think through the steps of identification.

TRACK IT ↘ Measure a Mammal Track

Measure an animal track and identify its maker.

WHAT YOU'LL NEED

- a ruler or measuring tape, a pencil or pen

STEP 1 Find a wild animal track. Look around; are there more of the same? Try to find both a front foot and hind foot track.

STEP 2 Draw a rough sketch of both tracks below.

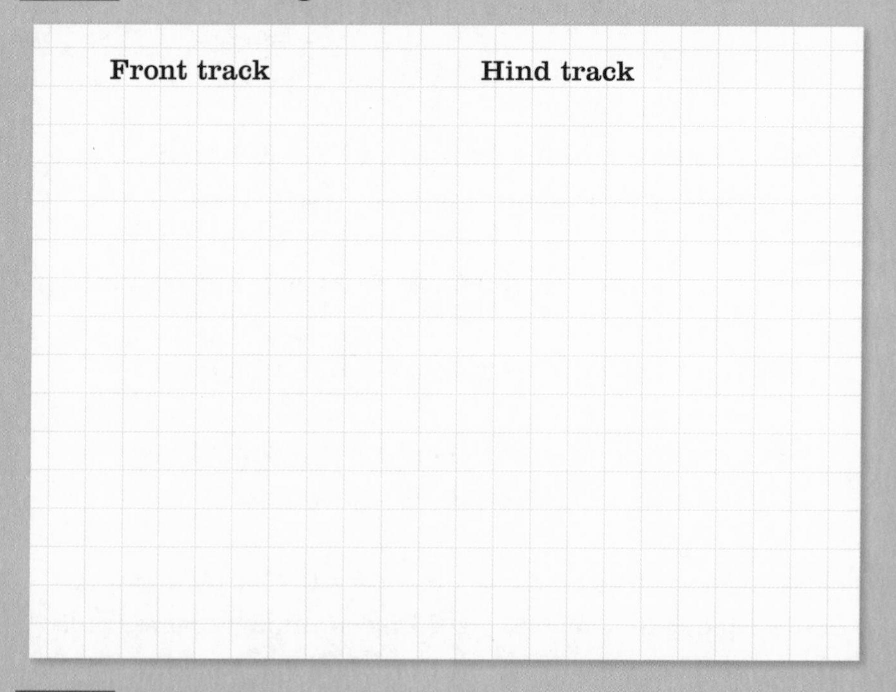

STEP 3 Use your ruler or the ruler on the back cover to measure the length and width.

STEP 4 Record the measurements, count the toepads, note any claws, and fill out the description below.

DATE

TIME

LOCATION

WEATHER

FRONT

Track length...............

Track width...............

Number of toes ....

☐ Claws ☐ No claws

HIND

Track length................

Track width ....

Number of toes ....

☐ Claws ☐ No claws

What else?

Who do you think made these tracks?

STEP 5 Use the Mammal Identification section (pages 240–301) and/or a field guide to help you identify the tracks.

TAKE IT TO THE NEXT LEVEL ↗

Measure a Stride

Find a set of mammal tracks with at least five prints that include all four feet in a row. Use a ruler or measuring tape to measure the animal's stride and straddle (see the diagram on page 217).

CHAPTER 4

How Mammals Live

You can find mammals everywhere! They live in all of North America's many habitats—from polar bears in frigid Alaska to kangaroo rats in California's scorching Death Valley. Many mammals are terrestrial, living on land and getting around on four legs. Others are aquatic, like dolphins and whales. These finned and flippered mammals breathe air, but are born in water and never leave. You'll have to go to the ocean to see them. Seals, otters, muskrats, and other semiaquatic mammals go back and forth between water and land.

BADGER

TERRESTRIAL

AQUATIC

DOLPHINS

SEA OTTER

SEMIAQUATIC

Bats are the only flying mammals, and are truly a sight to see. But many mammals spend time far from the ground up in trees. A variety of rodents including squirrels are arboreal tree dwellers, as well as opossums, porcupines, ringtail cats, and other expert climbers. Underground is another place mammals inhabit. Accomplished diggers like moles, badgers, and prairie dogs build homes, store food, or hunt those that do under the ground. What to spot them? Scout out their homes!

FLYERS

CLIMBERS

DIGGERS

Follow the Food

When watching mammals, think about how a squirrel or a fox is each adapted to its own particular way of life. All mammal bodies maintain a steady temperature by constantly burning fuel—food. Mammals spend a lot of time finding and eating food compared to snakes or frogs.

If you observe carefully, you'll notice that the bodies of mammals reflect *how* they make a living. Hooved mammals like sheep and deer, as well as many rodents, are plant-eating herbivores. Omnivores like raccoons, skunks, and bears eat whatever is around. Meat-eating carnivores include weasels, seals, dolphins, and cats, as well as insectivorous bats and shrews.

OMNIVORE

CARNIVORE

SHREW

INSECTIVORE

Predators, like cats and foxes, often have close-set eyes that look forward, creating binocular vision that better judges close distances, important for hunting. Pounce! Mammals who are preyed upon, like rabbits and deer, have eyes on opposite sides of their heads. They can look out for danger all around them. Many wear camouflage colors, too, and some are equipped with sophisticated defenses like quills (porcupine), armor (armadillo), and icky-smelling squirting liquid (skunks).

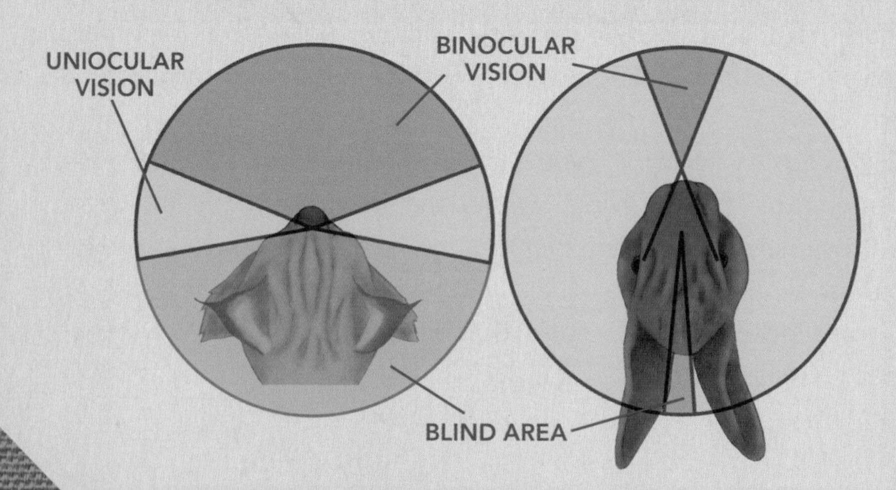

Teeth Make the Mammal

You can tell a lot from a mammal just by looking at its teeth—including whose mouth it came from! Teeth are a very specialized mammal body part. Grazing animals like elk have strong grinding teeth. Rodents have large front incisor teeth for gnawing that never stop growing. Carnivores like coyotes and otters have shredding teeth and piercing canines for holding on to prey. Like tools suited to a specific job, mammal teeth are made for crunching, grinding, slicing, or shredding the particular food of their owners. Chomp!

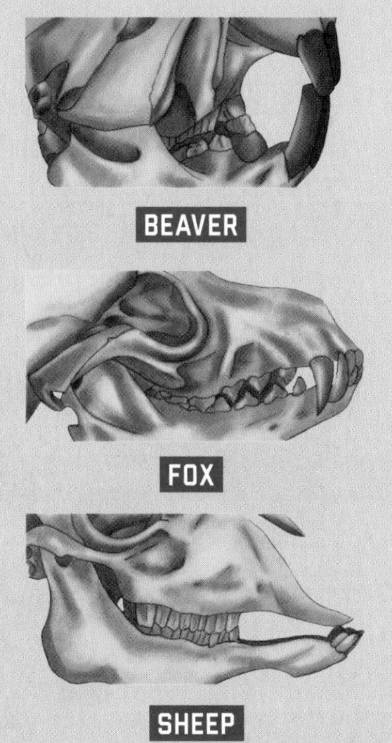

Everybody Poops

Mammal scat is another clue to who's around. The food that mammals shred, gulp, chomp, and swallow isn't on a one-way trip. What it looks like once it comes out the other end often gives away its owner. Some scat, like coyote or otter poop, is distinctive enough for species identification. Other kinds aren't quite as unique, but the shape, size, and texture tell you whether a mouse, bat, rabbit, or cat made it.

Scat can be easy to find. Wild mammals often want their poop to be noticed! It's a smelly way to say, *I'm here*. Deer leave scat along the trails they travel, and coyotes poop on people paths through parks. When you find some scat, notice its size, shape, color, and whether or not there are leftovers in it, like berry seeds, hair, or grass.

Here are some general guidelines:

HERBIVORE scat* is generally made up of many rounded pieces, called droppings, with plant fibers in them. The pieces get bigger with the animal. Moose droppings are grape-size, while those of rabbits are pea-size, for example.

CARNIVORE scat is often a single long piece that includes fur, bones, bug wings, or other remains of prey. It's often twisty with tapered ends.

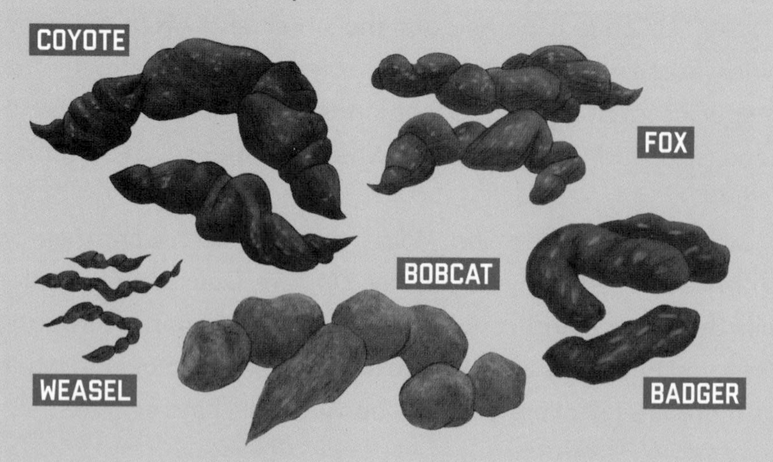

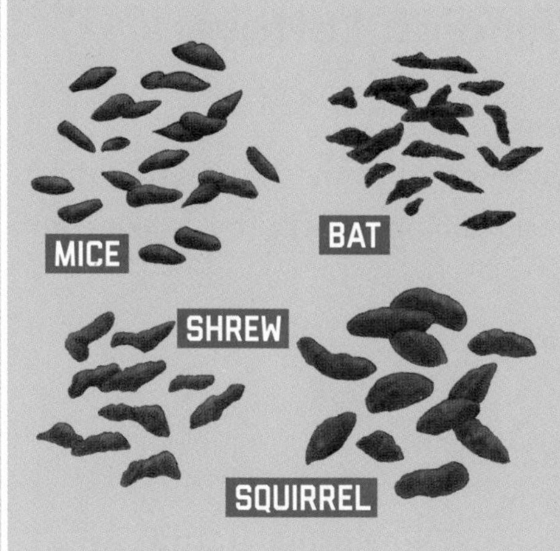

SMALL RODENT scat from mice, voles, and rats are rice-shaped pellets, whereas those of insect-eating shrews and bats are more shriveled versions with bug parts sometimes showing through.

OMNIVORE (like most of us!) scat often varies with the food that's been eaten. Raccoon, bear, and opossum scat usually comes in segments with blunt ends and can be full of seeds, half-digested berries, or whatever!

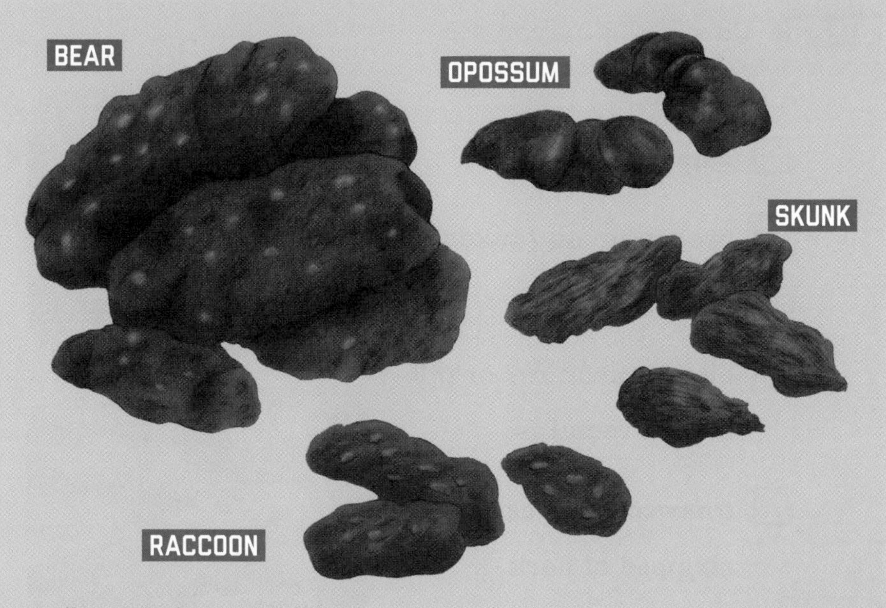

* All scat not drawn to scale.

TRY IT → Finding Leftovers

Mammals leave evidence lying around of what they've been eating. You can find clues about who's been dining nearby by looking for their leftovers.

WHAT YOU'LL NEED

- a pencil or pen

STEP 1 Go for a walk or hike in a park or other nature area. Along a creek, river, or the shore works, too.

STEP 2 Keep your eyes open for signs that mammals have eaten meals there recently. Look for evidence of teeth chewing, crunching, gnawing, clipping, or biting!

STEP 3 Check off any you see:

- [ ] emptied nutshells
- [ ] broken-up and taken-apart pinecones
- [ ] bits of rabbit fur or pile of bird remains
- [ ] gnawed-on sticks stripped of bark

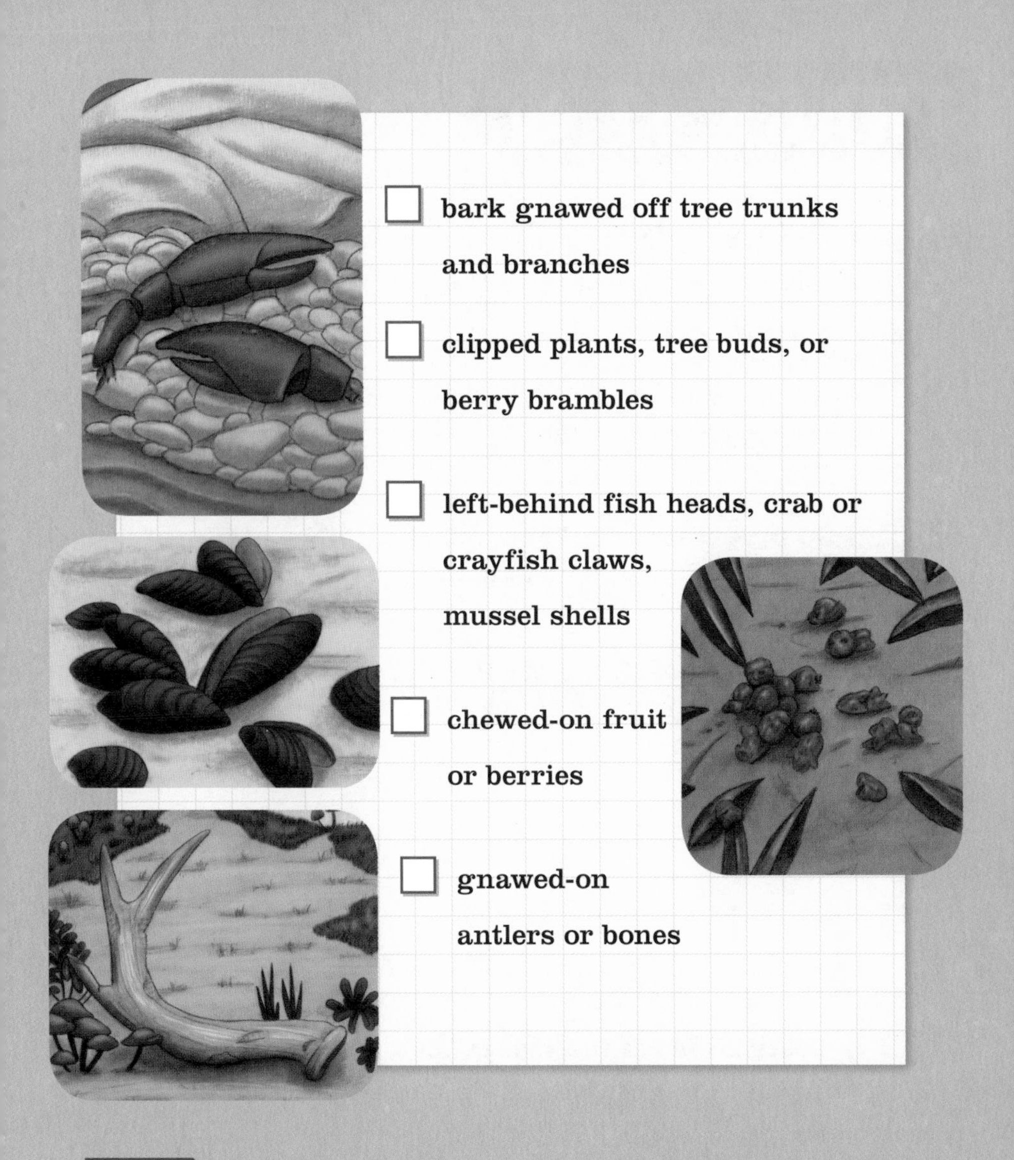

- [ ] bark gnawed off tree trunks and branches
- [ ] clipped plants, tree buds, or berry brambles
- [ ] left-behind fish heads, crab or crayfish claws, mussel shells
- [ ] chewed-on fruit or berries
- [ ] gnawed-on antlers or bones

STEP 4 Write in any ideas about whose leftovers you spotted. And if you touched anything, wash your hands. Critters can carry germs.

TRACK IT ↘ Go Scat Scouting

See what scat you can find! But don't touch without wearing gloves. It is poop, after all, and can carry germs.

WHAT YOU'LL NEED

- gloves, a flashlight, a hand lens or magnifying glass (optional), a pencil or pen

STEP 1 Go for a walk or hike in a park or other nature area. Look for animal-made trails and runways.

STEP 2 Search for scat on the ground, on fallen trees, and in grassy areas. Found some? Lucky you! Describe it below.

STEP 3 Describe where the scat is (on a stump one foot off the ground, on the ground next to a fence post, in the trail, next to the pond, etc.)

STEP 4 Look closely at these scatological characteristics:

- [ ] Size (inch-long, pea-size, etc.)
- [ ] Shape (round, oval, segmented, twisty, tapered, etc.)
- [ ] Color (all brown, brown and gray, etc.)
- [ ] Texture, anything recognizable in it? (fur, seeds, berries, insects, etc.)

STEP 5 Any ideas about whose scat you spotted? Write the name of the mammal. ______

Congrats! You're a scatology expert!

TAKE IT TO THE NEXT LEVEL

The Tales Teeth Tell

Find a skull or jawbone? (Wandering a bit off trail will up your odds.) When examining it, pay special attention to the teeth. Can you tell if the animal was a gnawer, grazer, or carnivore? Rodent skulls have big incisors on the top and bottom jaws, and then a gap before the cheek teeth. Carnivores have shredding triangular-shaped teeth in back, and long, pointy canines. Grazers like deer only have incisors on the bottom jaw, then a gap before thick, grinding cheek teeth.

CHAPTER 5

What Mammals Do

Spotting mammals, finding mammals' signs, and being able to identify them is awesome. What's even better? Observing mammals as they go about their lives. Watching a black bear mom help its cub down a tree, a beaver build a dam, or a bat hunt moths is as entertaining as any movie. Like us humans, other mammals care for young, have big brains, live in groups, use their senses to navigate the world, and interact with other species. All this adds up to all sorts of fascinating behavior. Ready to find out what mammals are up to? Then you'll be on your way to seeing some of it in action.

AMERICAN BEAVER

Birth and Growth

Mammals spend a lot of time caring for their young compared to other kinds of animals. The milk fed to offspring is a nutritious food that keeps young mammals with their mothers while learning how to survive on their own. Sometimes, as with mice and hares, the time between birth and weaning is only a few weeks. Many social mammals, including elephants and apes, nurse their young for a number of years, and the young stay with their mother or family group long after being weaned. This is the time when group members learn the rules, cooperation strategies, communication behaviors, and other how-tos of group living. Lucky enough to spot a baby mammal? Its mother and/or father are likely nearby and will defend the young. Back off, use binocs, and be aware. Don't get between parents and babies.

VIRGINIA OPOSSUM

Social vs. Solitary Mammals

Solitary mammals, such as pumas and bears, don't live in groups. But their mothers teach them to hunt and defend themselves before they strike out on their own. And solitary mammals also communicate to others of their kind. Leaving scat out in the open, scent marking territory, spraying urine, and scratching on trees or dens are all signposts that declare NO TRESPASSING.

Migration and Hibernation

Mammals need to eat regularly to fuel their bodies. Those that live in places where food isn't available year round must have a plan B. Insect-eating bats avoid winter starvation by either migrating south (Mexican free-tailed bats) or living on stored fat in hibernation until spring (little brown bats). Most rodents either hibernate come autumn (woodchucks), or stash food in underground burrows or hollow trees (squirrels) for winter snacking. If you're scouting for mammals to see, follow the food! Look for mammals where they forage or hunt.

LITTLE BROWN BATS

Using Their Senses

Many mammals use the same five senses that you do—touch, smell, taste, hearing, and sight. But whales can't smell much, and many mammals feel with whiskers, tails, and other body parts. Humans have a lousy sense of smell as far as mammals go. Detecting enemies, friends, and what's going on via odors is most mammals' best source of info. Hearing is another sense that is better in most other mammals than humans. You might not hear that chipmunk in the woods, but it probably hears you stomping along. Bats, dolphins, and some whales use echolocation to navigate and find food thanks to their superb hearing. Seeing is the superior sense for primates, including people. Few other mammals have color vision as good as humans'. Most mammal eyes are built to see movement over fine detail. That means if you're quiet, downwind, and standing still, a mammal might not see you watching it—at least for a while. Nice work!

EASTERN CHIPMUNK

The Only Exception—Bats!

Bats are the only winged mammal there is! And they're so fun to watch—they dive and swoop after flying insects on summer nights and can be found in the countryside and cities.

SILVER-HAIRED BAT

Here are some bat watching tips:

- The best time for seeing bats is just before sunset. Bats are starting to come out, but the sky is still light enough to see well.
- Areas with lots of sky can help you see bats more easily. Near rivers, lakes, parks and ball fields, forest edges, open hilltops, etc.
- Good bat watching places have lots of tasty beetles, moths, flies, and mosquitoes for these flying mammal insectivores.
- City streetlights and athletic field lights attract insects, which can bring in bats. That's an easy place to look, even well after dark.
- Look nearby for known bat roosting sites, like under a bridge, in an old barn, or at a frequented bat house.
- If you live in the desert Southwest, look for nectar-eating bats near flower gardens and hummingbird feeders at night.
- Bat sounds: You might hear bats making squeaky communication sounds—*Get out of my way. Hey, that was my moth*, etc. Echolocation sounds are ultrasonic and too high for humans to hear.
- Bat manners: Do not disturb hibernating bats in winter. Waking up uses a lot of stored fat and can cause a bat to starve before spring. Don't shine flashlights at sleeping, roosting bats. It's rude!
- Bat safety: NEVER touch a bat. Like raccoons, foxes, and many other wild mammals, bats can carry rabies.

TRY IT → Watch Mammals in Action

Can you find some mammals behaving like, well, *animals*? Review the Mammal Scouting Tips on page 194 and head out. Good luck!

WHAT YOU'LL NEED

- binocs

STEP 1 Go for a walk or hike in a park or other nature area. Woods, prairie edges, mountains, and along creeks and other waterways are all good.

STEP 2 Keep your eyes open and scan for mammals living there. Look up and down!

Woodchuck making a burrow.

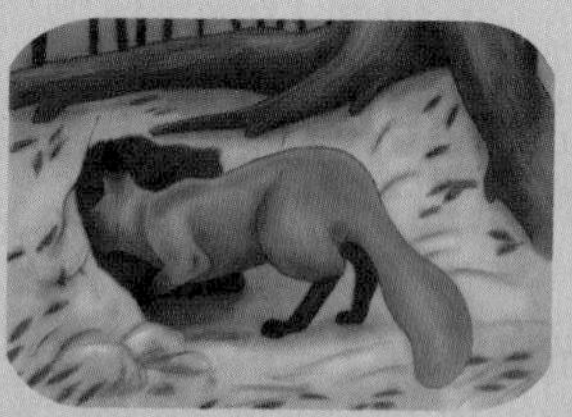

Fox entering its den.

Deer fawn sleeping while hidden.

Mole digging.

Tree squirrel building a nest.

Deer scraping tree with antlers.

STEP 3 When you see some action, settle in and observe what's going on. Use the TRACK IT below to record what you saw. Now that's wildlife watching!

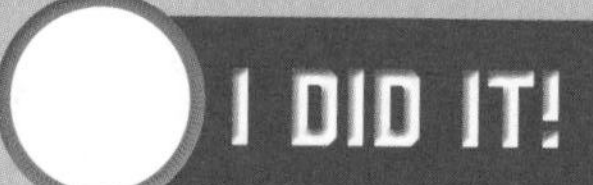

TRACK IT ↘ Record Mammal Behavior

Beaver cutting down a tree.

Wildlife scientists keep detailed notes of the animal behavior they observe. You can, too!

WHAT YOU'LL NEED

- a pencil or pen

STEP 1 Record what you observed a mammal doing by filling in the page below.

STEP 2 Did you recognize the mammal? Look for it in the Mammal Identification section (pages 240–301). Make a positive ID? Check off its I SAW IT! box and fill in the blanks.

DATE

TIME

LOCATION

WEATHER

What sort of mammal was it (chipmunk, squirrel, mouse, deer, etc.)?

What was it doing?

Describe the behavior.

How long did the behavior last?

I DID IT! DATE:

TAKE IT TO THE NEXT LEVEL ↗

Follow the Trail

Mammals move around to find what they need. Can you figure out why?

STEP 1 Look for and locate a mammal trail. Common trails to look for are: a deer-hoof dotted trail through the woods, a path to water, or a runway tunnel through grass.

STEP 2 Follow the trail as far as it goes (or as far as you can!) without stopping.

STEP 3 Look for any animal tracks you see. Try to identify them. Are they all going in a single direction, or back and forth?

STEP 4 Walk around the trail you found. Can you tell where the mammal is going? Look for clues. Is there water, food, or shelter nearby?

I DID IT! DATE:

WHITE-TAILED DEER

TRY IT → Watch Some Fur Fly

Go bat watching on a summer evening. (You might want to take a second look at the bat watching tips on page 235.)

LITTLE BROWN BATS

WHAT YOU'LL NEED

- a strong flashlight

STEP 1 Head out just before sunset and search for places with yummy flying beetles and moths where bats hunt. Like where? Spots near rivers and lakes, forest edges, parks and ball fields, or known bat roosting sites, like near old bridges or barns.

STEP 2 Look in the sky for fluttering fliers with upturned wings.

STEP 3 Once it's totally dark, shine the flashlight around tree canopies to look for bats flying.

MAMMAL IDENTIFICATION

Welcome to your guide to mammal identification. Here are some tips to get started:

Mammals are organized by size, from large to medium to small. Large includes hoofed animals, big predators like foxes and bears, seals, whales, etc. Medium mammals range from possums and beavers to weasels. Small mammals are everyone else—mice, moles, chipmunks, squirrels, etc.

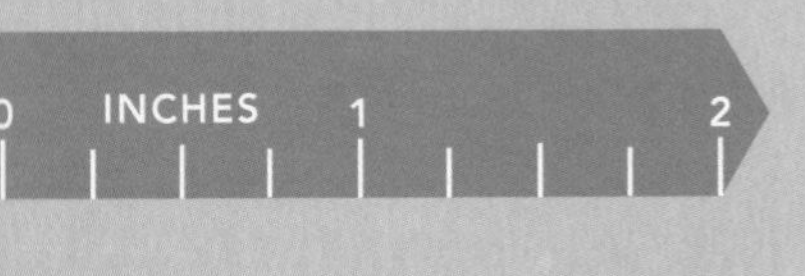

Note that lengths list the body and tail separately.

The widths of tracks are averages; they can be smaller or larger depending on the individual animal, as well as whether the track is found in snow, sand, or mud. Animals can sometimes leave incomplete tracks. The part of the foot that doesn't leave a track is shown as an outline.

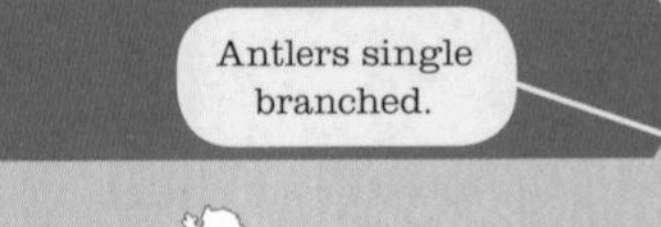

The labeled main illustration points out field marks. Take note!

Check the range map to see where the mammal lives.

SNOUT OR MUZZLE
EAR
TAIL
CHEST
FRONT LEG
PAW
BELLY
DOUGLAS SQUIRREL
HIND LEG
RUMP

LARGE MAMMALS

White-tailed Deer

(Odocoileus virginianus)

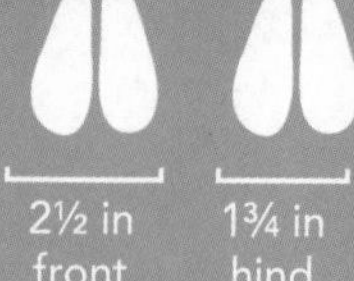

SIZE 24–42 inches (61–106½ cm) in height at shoulder, males larger

SHAPE Slim deer with medium-size ears, a long, full tail, and large eyes. Males have paired antlers, each with a single main branch.

COLOR Has a gray coat in the winter and a rusty brown coat in the summer. It has a white belly, inner thighs, chin, throat, and underside of its tail. Fawns are reddish brown with white spots.

BEHAVIOR Nocturnal or crepuscular. When alarmed, it snorts and raises its tail as it bounds off. It lives in small groups of females and young, or all males.

HABITAT Forest edges, river corridors, brushland, farmland, parks.

POINT OF FACT It makes beds of matted down grass, leaves, or snow to rest. Males leave small trees scraped up from their antlers, too.

I SAW IT!

WHEN I SAW IT
DATE

WHERE I SAW IT
SPECIFIC LOCATION, INCLUDE STATE

WHAT IT WAS DOING

NOTES

Mule Deer

(Odocoileus hemionus)

2½ in front

1¾ in hind

SIZE 36–42 inches (91½–106½ cm) in height at shoulder, males larger

SHAPE Medium-built deer with large ears, a short, skinny tail, and large eyes. Males have paired antlers, each dividing into two branches.

COLOR Has a gray coat in the winter and a brown coat in the summer. It has a white belly, inside thighs, chin, throat, and rump patch with a black-tipped tail. Fawns are spotted.

BEHAVIOR Nocturnal or crepuscular. When alarmed, it bounces away with all four of its hoofs hitting the ground at the same time.

HABITAT Desert, brushland, forest, and mountains.

POINT OF FACT The four-footed hopping movement is called *stotting*. White-tailed deer don't hop this way.

I SAW IT!

WHEN I SAW IT
DATE

WHERE I SAW IT
SPECIFIC LOCATION, INCLUDE STATE

WHAT IT WAS DOING

NOTES

LARGE MAMMALS

American Elk or Wapiti

(Cervus canadensis)

$3^{5}/_{8}$ in front | $3^{3}/_{16}$ in hind

SIZE 48–60 inches (122–152½ cm) in height at shoulder, males larger

SHAPE Large deer with a thick neck with shaggy fur, small ears, and a short tail. Males have antlers.

COLOR Dark brown head, neck, and legs. It has a lighter back, sides, and belly, and a cream rump patch and tail.

BEHAVIOR Mainly nocturnal but also crepuscular. It lives in large herds that move seasonally by the foods that are available.

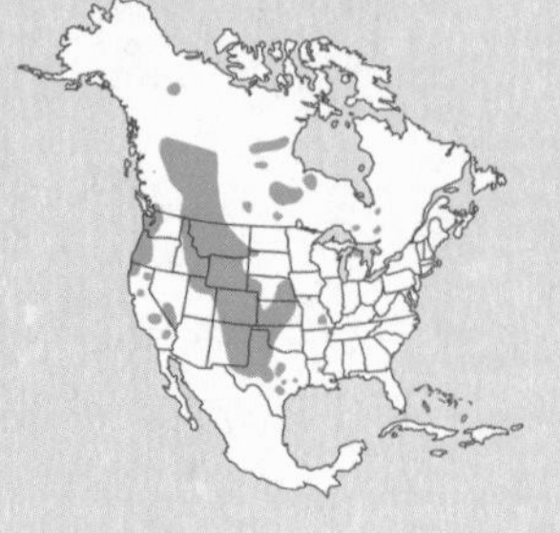

HABITAT Woods, meadows, low plains, and high mountains.

POINT OF FACT Wapiti comes from the name for the animal in the Shawnee language, *wapiti*, which means "white rump." Eurasian elk (*Alces alces*, page 246) are called moose in North America. It's confusing!

I SAW IT!

WHEN I SAW IT
DATE

WHERE I SAW IT
SPECIFIC LOCATION, INCLUDE STATE

WHAT IT WAS DOING

NOTES

LARGE MAMMALS

Bighorn Sheep

(Ovis canadensis)

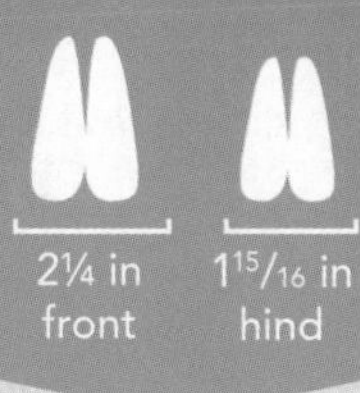

SIZE 36–42 inches (91½–106½ cm) in height at shoulder, males much larger

SHAPE Lean sheep with a tiny tail, long legs, and horns. Female horns are small, narrow, and are curved slightly backward. Male horns are much larger coils.

COLOR Yellow to gray brown fur on its body with a white muzzle, belly, and rump patch.

BEHAVIOR Diurnal and social. Rams famously butt their heads, clashing their horns in contest over females. It lives in herds separated by sex.

HABITAT Rugged, steep, mountain areas, high meadows, and deserts.

POINT OF FACT It's an agile climber that can walk along ledges as narrow as 2 inches (5 cm) wide.

I SAW IT!

WHEN I SAW IT

DATE

WHERE I SAW IT

SPECIFIC LOCATION, INCLUDE STATE

WHAT IT WAS DOING

NOTES

LARGE MAMMALS

Moose

(Alces alces)

5 in front

4 in hind

Antlers.

Nose.

Loose fold of skin, or dewlap.

MALE

FEMALE

SIZE 78–90 inches (198–228½ cm) in height at shoulder, males larger

SHAPE Enormous deer with shoulders higher than its rump, a long face with a huge nose, long ears, a chin dewlap, and a short tail stump. Males have heavy, large, and wide antlers.

COLOR Black-brown with lighter brown legs. Young are a red-brown.

BEHAVIOR Mostly crepuscular and usually solitary. It spends time feeding near water.

HABITAT Northern forest, tundra, willow thickets, swamps.

POINT OF FACT It's the largest kind of deer in North America. A male bull can weigh up to 1,800 pounds (816 kg).

I SAW IT!

WHEN I SAW IT
DATE

WHERE I SAW IT
SPECIFIC LOCATION, INCLUDE STATE

WHAT IT WAS DOING

NOTES

LARGE MAMMALS

Pronghorn

(Antilocapra americana)

MALE

1¾ in front

1 13/16 in hind

SIZE 30–42 inches (76–106½ cm) in height at shoulder, males a bit larger

SHAPE Medium-size, hooved, horned mammal with a wide snout, a thick neck, thin legs, medium-length ears, and a stubby tail. Female antlers are shorter than males and are not pronged.

COLOR Rust to light tan with a white belly, lower sides, and rump patch, bands on its cheeks and neck, and black horns. Males have black and white markings and have manes.

BEHAVIOR Diurnal or nocturnal. When alarmed, it flashes and fans its long hairs on its white rump.

HABITAT Grasslands, sagebrush, desert.

POINT OF FACT It's the fastest mammal in North America. It cruises at 25 miles (40 km) per hour and can sprint up to 70 miles (112 km) per hour.

WHEN I SAW IT

DATE

WHERE I SAW IT

SPECIFIC LOCATION, INCLUDE STATE

WHAT IT WAS DOING

NOTES

LARGE MAMMALS

Black Bear

(Ursus americanus)

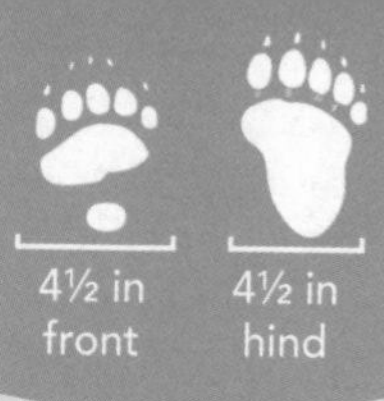

SIZE 48-72 inches (122–183 cm) in body length, 100–900 pounds (45–408 kg), males larger

SHAPE Medium to large bear with a rump higher than its shoulders, a straight snout, small eyes, and round ears.

COLOR Varies, but it usually has a black body with a brown muzzle. It can also be blond, rusty brown, or dark brown.

BEHAVIOR Diurnal and mostly solitary. It shuffles while foraging and can climb well.

HABITAT Forests and swamps, mountains, tundra.

POINT OF FACT If threatened, it will huff and gallop into a sprint of up to 35 miles (56 km) per hour.

I SAW IT!

WHEN I SAW IT

DATE

WHERE I SAW IT

SPECIFIC LOCATION, INCLUDE STATE

WHAT IT WAS DOING

NOTES

....

....

LARGE MAMMALS

Red Fox

(Vulpes vulpes)

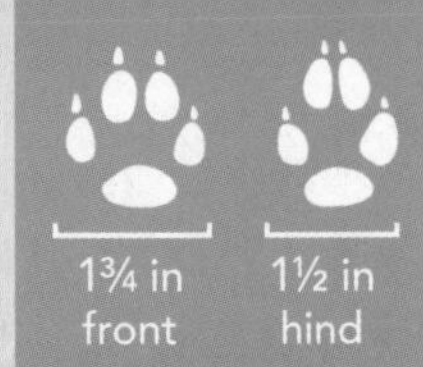

SIZE 17–20 inches (43–51 cm) in body length, plus 10–14 inches (25½–35½ cm) of tail

SHAPE Medium-size fox with long legs, a long, bushy tail, and triangle-shaped ears.

COLOR Varies, but usually an orange-red with black legs and ears, and a white belly, throat, chin and tail tip. The young, called kits, are medium brown.

BEHAVIOR Nocturnal or crepuscular. It moves and hunts like a cat by stalking and pouncing. It has high-pitched whines, yaps, and barks.

HABITAT Forest edges, fields, parks, marshes, neighborhoods.

POINT OF FACT Families stay together: pairs mate for life and both help raise kits. Daughters sometimes help with the next litter, too.

I SAW IT!

WHEN I SAW IT

DATE

WHERE I SAW IT

SPECIFIC LOCATION, INCLUDE STATE

WHAT IT WAS DOING

NOTES

LARGE MAMMALS

Gray Fox

(Urocyon cinereoargenteus)

SIZE 19–28 inches (48½–71 cm) in body length, plus 11–17 inches (28–43 cm) of tail

SHAPE Medium-size fox with a long, bushy tail, medium-length legs, and triangle-shaped ears.

COLOR Grizzled gray back and top of its tail, rusty tan ears, sides, legs, and bottom of its tail, a white belly and throat, and a black tail tip. Kits are medium brown.

BEHAVIOR Nocturnal or crepuscular. It easily climbs up trees and on branches to escape predators.

HABITAT Deciduous forests, old fields in East, rugged riverside forests in West.

POINT OF FACT It has sharp, curved claws that help it to climb up tree trunks. It also jumps from branch to branch!

I SAW IT!

WHEN I SAW IT

DATE

WHERE I SAW IT

SPECIFIC LOCATION, INCLUDE STATE

WHAT IT WAS DOING

NOTES

LARGE MAMMALS

Coyote

(Canis latrans)

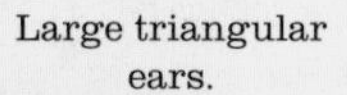

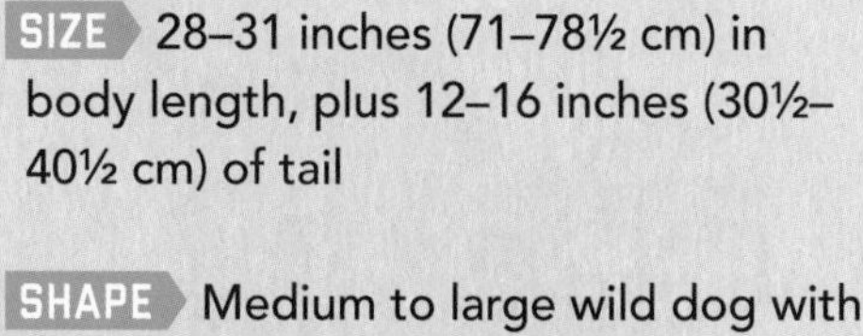

SIZE 28–31 inches (71–78½ cm) in body length, plus 12–16 inches (30½–40½ cm) of tail

SHAPE Medium to large wild dog with long legs, tall, pointy, large ears, a narrow snout, and a bushy tail.

COLOR Varies, but usually gray to tan with rust on its face and legs. Its eyeshine is a green-yellow. Pups are medium brown.

BEHAVIOR Mostly crepuscular. It runs with its tail held down and makes short, high yips and screaming, frantic howls. It usually travels alone or in pairs.

HABITAT Desert, grassland, forest edges, parks, towns, cities.

POINT OF FACT It's extremely adaptable and omnivorous! It hunts everything from birds and snakes to rabbits and deer. It also eats bugs, carrion, fruit, and garbage. Some are solitary and others form packs.

I SAW IT!

WHEN I SAW IT
DATE

WHERE I SAW IT
SPECIFIC LOCATION, INCLUDE STATE

WHAT IT WAS DOING

NOTES

LARGE MAMMALS

Bobcat

(Lynx rufus)

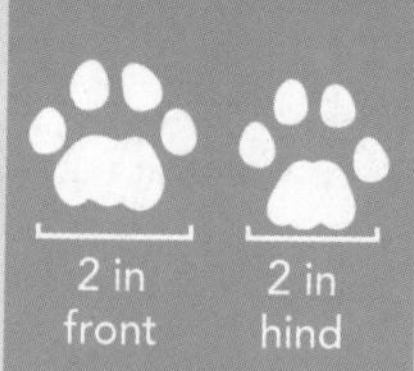

SIZE 20–39 inches (51–99 cm) in body length, plus 4–8 inches (10–20½ cm) of tail

SHAPE Small, stocky wildcat with longish legs, ears with tiny tufts, a wide face, and a very short tail.

COLOR Colors and spots vary from light tan and rusty brown, to gray and black. Spots and bars can either be sharply defined or more faded.

BEHAVIOR Mostly nocturnal and mainly solitary. It can swim and climb well, but it travels by walking. It stalks its prey on the ground and then pounces.

HABITAT Forests, swamps, mountains, grasslands.

POINT OF FACT Its loud cry, or caterwaul, during mating season sounds like a screaming person or crying baby.

I SAW IT!

WHEN I SAW IT

DATE

WHERE I SAW IT

SPECIFIC LOCATION, INCLUDE STATE

WHAT IT WAS DOING

NOTES

Mountain Lion

(Puma concolor)

3½ in front | 3¼ in hind

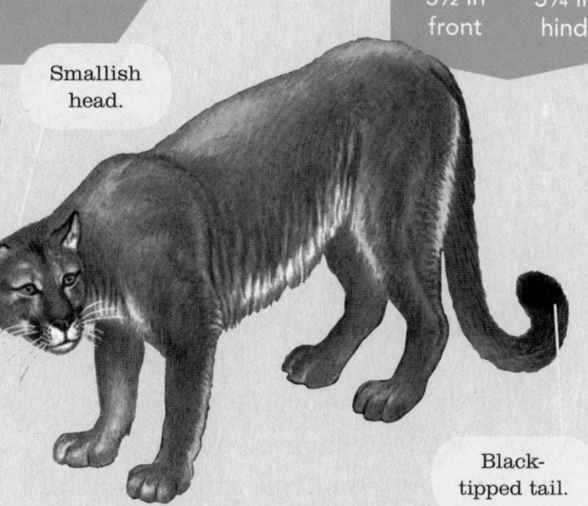

SIZE 36-60 inches (91½–152½ cm) in body length, plus 24–40 inches (61–101½ cm) of tail

SHAPE Very large wild cat with a small head and a long tail.

COLOR Sandy to red-brown with a cream belly and chest, and a black-tipped tail. Cubs are spotted.

BEHAVIOR Both diurnal and nocturnal. It's usually solitary and travels long distances within huge territories.

HABITAT Deserts, forests, mountains, swamps.

POINT OF FACT It lives anywhere from Chile to Canada, and you may know it by any of its many names: puma, cougar, mountain lion, catamount, panther, or painter.

I SAW IT!

WHEN I SAW IT
DATE

WHERE I SAW IT
SPECIFIC LOCATION, INCLUDE STATE

WHAT IT WAS DOING

NOTES

Harbor Seal

(Phoca vitulina)

SIZE 48-72 inches (122–183 cm) in length, 100–370 pounds (45–168 kg), males larger

SHAPE Medium-size seal with no ears on its head, a short snout with nostrils in a V shape, large eyes, and small front flippers.

COLOR Varies, but usually it is white to yellow with dark spots all over. It sometimes has dark rings and a pale muzzle and belly.

BEHAVIOR It's very social and basks on rocks with others. It scoots around on land and often holds up its rear flippers and head in a banana-like posture.

HABITAT Coastal waters, rocky shores, and some estuaries, rivers, and lakes.

POINT OF FACT It's a hunter of fish, squid, clams, and shrimp. It can dive up to 1,500 feet (457 m), and stay underwater for up to half an hour.

I SAW IT!

WHEN I SAW IT
DATE

WHERE I SAW IT
SPECIFIC LOCATION, INCLUDE STATE

WHAT IT WAS DOING

NOTES

....

....

LARGE MAMMALS

California Sea Lion

(Zalophus californianus)

Ear flaps.

Upright posture propped on flippers.

SIZE 60-96 inches (152½–244 cm) in length, 110–900 pounds (50–408 kg), male much larger

SHAPE Large, sleek seal with a visible neck and a dog-like head with small ears, large eyes, and thick whiskers. Its large back flippers turn forward for walking.

COLOR Yellow-brown to black when wet.

BEHAVIOR It's social and noisy. Groups haul out onto land and bark, growl, and roar. It gallops on land by using its flippers.

HABITAT Coastal ocean waters and small islands.

POINT OF FACT It's a superb swimmer that dives to depths of 450 feet (137 m) and can stay underwater for twenty minutes.

WHEN I SAW IT

DATE

WHERE I SAW IT

SPECIFIC LOCATION, INCLUDE STATE

WHAT IT WAS DOING

NOTES

....

....

LARGE MAMMALS

Bottlenose Dolphin

(Tursiops truncatus)

Tall and curved dorsal fin.

SIZE 6–13 feet (183–396 cm) in length, 440–600 pounds (200–272 kg)

SHAPE Large, thick-bodied dolphin with a short, beaked snout and a tall, curving dorsal fin.

Short, thick, beaked snout.

COLOR Shiny-gray skin with a lighter colored underbelly.

BEHAVIOR Lives in groups of 10–25 dolphins. It surfs waves alongside ships and leaps out of the water.

HABITAT Coastal and offshore warmer waters.

POINT OF FACT It communicates with both body language and a series of squeaks and whistles from its blowhole. It also uses clicks as a kind of sonar to find prey underwater.

I SAW IT!

WHEN I SAW IT

DATE

WHERE I SAW IT

SPECIFIC LOCATION, INCLUDE STATE

WHAT IT WAS DOING

NOTES

LARGE MAMMALS

Gray Whale

(Eschrichtius robustus)

SIZE 36–50 feet (11–15 m) in length, 16–33 tons (15–30 metric tons)

SHAPE Large, long whale with a narrow head and a huge mouth, barnacles on its head and upper back, lobed flippers, and ridges down its lower back to its tail.

COLOR Blotchy gray skin with white barnacle patches.

BEHAVIOR Blows out air 3–5 times before diving. It raises its head vertically out of the water to look around.

HABITAT Coastal ocean waters.

POINT OF FACT It eats by vacuuming up invertebrates from the sea floor and filtering them out with flexible, comb-like structures in its mouth, called baleen plates.

I SAW IT!

WHEN I SAW IT
DATE

WHERE I SAW IT
SPECIFIC LOCATION, INCLUDE STATE

WHAT IT WAS DOING

NOTES

LARGE MAMMALS

Humpback Whale

(Megaptera novaeangliae)

SIZE 36–49 feet (11–15 m) in length, 25–44 tons (23–40 metric tons)

SHAPE
Large, curve-backed whale with a wide head, a huge mouth, long, thin flippers, and a small dorsal fin.

COLOR Dark gray skin with some white on its flippers, belly, throat, and tail.

BEHAVIOR Blows out air 2–8 times before diving. It can leap out of the water when breaching, and it slaps the water with its tail or flippers.

HABITAT Coastal areas or near islands in winter and summer, open ocean during spring and autumn migration.

POINT OF FACT Males sing long, complicated songs underwater during their breeding season.

I SAW IT!

WHEN I SAW IT

DATE

WHERE I SAW IT

SPECIFIC LOCATION, INCLUDE STATE

WHAT IT WAS DOING

NOTES

MEDIUM-SIZE MAMMALS

American Badger

(Taxidea taxus)

2 in front | 1¾ in hind

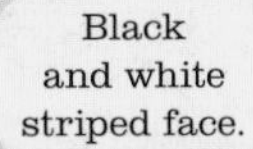

SIZE 20–26 inches (51–66 cm) in body length, plus 4–6 inches (10–15 cm) of tail

SHAPE Stocky, short-legged, beefy weasel with a wide head, small ears, and a short tail.

COLOR Gray to red long fur on its back and sides, a lighter, buff belly, and dark feet. It has a white throat and chin with black face patches and a white strip from its nose to its shoulders.

BEHAVIOR It's mostly nocturnal, but a badger out during the day might be a hunting mom who spends nights with her kits.

HABITAT Prairies, mountain meadows, deserts, open marshland, and grasslands.

POINT OF FACT It digs deep, tunneled dens to raise its young, as well as shallower burrows to rest in, and hunting burrows to hide itself and surprise-attack passing prey.

WHEN I SAW IT

DATE

WHERE I SAW IT

SPECIFIC LOCATION, INCLUDE STATE

WHAT IT WAS DOING

NOTES

MEDIUM-SIZE MAMMALS

Raccoon

(Procyon lotor)

2 in front

1¾ in hind

SIZE 16–24 inches (40½–61 cm) in body length, plus 6–16 inches (15–40½ cm) of tail

SHAPE Small, bear-like body with short legs, a small head with ears on top, and a long, bushy tail.

COLOR Gray-brown body fur, a black nose, an eye mask (black), and white muzzle sides and eyebrows. Its tail has bands of cream and black, and it has a yellow eyeshine.

BEHAVIOR Mostly nocturnal. It walks quickly with its back arched and its head low. It can climb with ease.

HABITAT Wetlands, woods, cities, neighborhoods.

POINT OF FACT It makes a variety of sounds, including aggressive screams, growls, and squeals. Moms trill to their young.

I SAW IT!

WHEN I SAW IT

DATE

WHERE I SAW IT

SPECIFIC LOCATION, INCLUDE STATE

WHAT IT WAS DOING

NOTES

....

....

Virginia Opossum

(Didelphis virginiana)

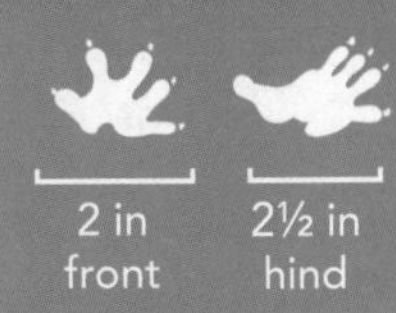

SIZE 14–20 inches (35½–51 cm) in body length, plus 9–15 inches (23–38 cm) of tail

SHAPE Large, scruffy, ratlike body with a long, naked, scaly tail, deep-set eyes, a pointy snout, and rounded ears.

COLOR Grizzled gray and brown fur on its body, a white head, dark ears, and a pink nose and toes. Its eyeshine is bright red.

BEHAVIOR Nocturnal. It waddles when it walks on the ground while holding its tail up, and is also an expert climber. It will play dead if it's cornered by lying on its side, drooling, and sticking its tongue out.

HABITAT Woods, neighborhoods, old fields, roadsides.

POINT OF FACT It's the only native marsupial (pouched mammals like kangaroos) in North America. It gives birth to 8–16 tiny young that crawl into their mother's pouch, attach to a nipple, and grow there for two months.

I SAW IT!

WHEN I SAW IT

DATE

WHERE I SAW IT

SPECIFIC LOCATION, INCLUDE STATE

WHAT IT WAS DOING

NOTES

....

....

MEDIUM-SIZE MAMMALS

American Mink

(Neovison vison)

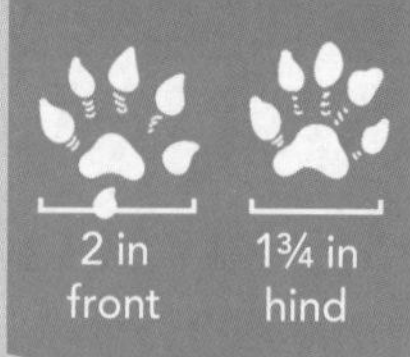

SIZE 12–16 inches (30½–40½ cm) in body length, plus 6–8 inches (15–20½ cm) of tail

SHAPE Medium-size weasel with a long, low body and a long, furry tail.

COLOR Dark brown body with white on its chin and chest, and a black tail tip. Its eyeshine is yellow-green.

BEHAVIOR Mostly nocturnal or crepuscular. It's a terrific swimmer with semi-webbed hind toes.

HABITAT Along wooded streams, lakes, marshes, and swamps.

POINT OF FACT It purrs when it's happy, hisses and snarls when it's alarmed, and sometimes chuckles, too.

I SAW IT!

WHEN I SAW IT
DATE

WHERE I SAW IT
SPECIFIC LOCATION, INCLUDE STATE

WHAT IT WAS DOING

NOTES

Ringtail

(Bassariscus astutus)

SIZE 13–15 inches (33–38 cm) in body length, plus 13–16 inches (33–41 cm) of tail

SHAPE Slim, cat-like shape with a bushy tail as long as its body, large eyes, and a short, pointed snout on its face.

COLOR Brown-gray body, white fur around its eyes, and black and white bands on its tail. Its eyeshine is a bright red-orange.

BEHAVIOR Nocturnal. Its unique wrists can swivel backward, helping it to scale canyon walls and slide quickly along tree branches with smooth, gliding movements.

HABITAT Dry, rocky mountain and canyon regions of the West.

POINT OF FACT It's also known as the "miner's cat" because it was welcome in mining camps as an excellent mouse hunter.

I SAW IT!

WHEN I SAW IT

DATE

WHERE I SAW IT

SPECIFIC LOCATION, INCLUDE STATE

WHAT IT WAS DOING

NOTES

MEDIUM-SIZE MAMMALS

Western Spotted Skunk

(Spilogale gracilis)

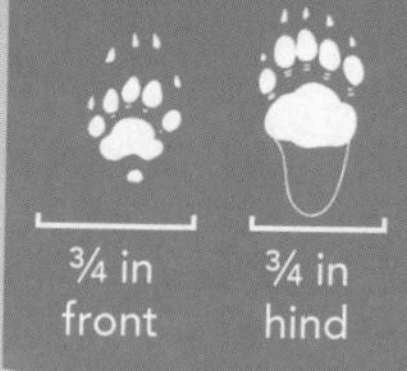

SIZE 8–11 inches (20½–28 cm) in body length, plus 4½–6 inches (11½–15 cm) of tail

SHAPE Small skunk with large, round ears and a short, bushy tail.

COLOR Black with rows of white spots and stripes on its body, and a white triangle in between its eyes.

BEHAVIOR Nocturnal. It's fast and able to climb trees. When it's alarmed, it turns around and does a handstand to aim musk glands and tail toward its attacker, then sprays if pushed further.

HABITAT Open woodlands, farmlands, canyons in West.

POINT OF FACT Its cousin, the eastern spotted skunk (*Spilogale putorius*) has somewhat narrower white bands, a mostly black tail, and lives east of the Rocky Mountains. But they're hard to tell apart!

I SAW IT!

WHEN I SAW IT

DATE

WHERE I SAW IT

SPECIFIC LOCATION, INCLUDE STATE

WHAT IT WAS DOING

NOTES

MEDIUM-SIZE MAMMALS

North American Porcupine

(Erethizon dorsatum)

1½ in front | 1¾ in hind

SIZE 24–36 inches (61–91½ cm) in body length, plus 6–12 inches (15–30½ cm) of tail

SHAPE Large, round, tubby rodent that's covered in quills and long hair with small eyes, a round snout, and a medium-length tail.

COLOR Yellow quills on its head, rump, and tail. Its long hairs on its back and shoulders are brown to tan, and it has a brown face.

BEHAVIOR Mostly nocturnal. It ambles slowly on the ground and hoists itself up onto trees. When threatened, it turns around and flicks tail.

HABITAT Deciduous and coniferous forests.

POINT OF FACT It can't throw its quills, but whatever it whacks with its tail gets stuck with them. The quill ends have barbs that make them hard to pull out.

WHEN I SAW IT

DATE

WHERE I SAW IT

SPECIFIC LOCATION, INCLUDE STATE

WHAT IT WAS DOING

NOTES

....

....

MEDIUM-SIZE MAMMALS

Striped Skunk

(Mephitis mephitis)

1 in front

1 in hind

White stripe on face.

Stripes down back.

SIZE 12–20 inches (30½–51 cm) in body length, plus 9–14 inches (23–35½ cm) of tail

SHAPE Medium-size, husky skunk with a bushy tail just shorter than its body length and small ears.

COLOR Black with thin, white stripe on its face, one or two wide, white stripes down its back, and a black and white tail.

BEHAVIOR Nocturnal. It rummages through leaf litter and walks slowly. If alarmed, it stops and stamps its front feet while hissing, then puts up its tail and sprays if pushed.

HABITAT Fields, open woods, neighborhoods, deserts.

POINT OF FACT Its size can vary: It's bigger in northern states than in southern states, and males are larger than females. Its stripe pattern on its back can vary, too: It can have one thick stripe or two.

I SAW IT!

WHEN I SAW IT

DATE

WHERE I SAW IT

SPECIFIC LOCATION, INCLUDE STATE

WHAT IT WAS DOING

NOTES

MEDIUM-SIZE MAMMALS

Black-tailed Jackrabbit

(Lepus californicus)

1¼ in front

1½ in hind

SIZE 17–20 inches (43–51 cm) in body length, plus 2½–4½ inches (6½–11½ cm) of tail

SHAPE Big, lean hare with long legs and feet, very long ears, big eyes, and a medium-length tail.

COLOR Light brown and gray back and sides with a white belly, black-tipped ears, and a tail with black on top and a light color below.

BEHAVIOR Nocturnal or crepuscular. It's a fast and agile runner, and it thumps its hind feet when alarmed.

HABITAT Prairies, farmland, pastures, deserts.

POINT OF FACT It runs at speeds up to 35 miles (56 km) per hour and leaps as high as 20 feet (6 m).

WHEN I SAW IT

DATE

WHERE I SAW IT

SPECIFIC LOCATION, INCLUDE STATE

WHAT IT WAS DOING

NOTES

MEDIUM-SIZE MAMMALS

Eastern Cottontail

(Sylvilagus floridanus)

SIZE 12–16 inches (30–40½ cm) in body length, plus 1½–3 inches (4–7½ cm) of tail

SHAPE Medium-size, stocky rabbit with big eyes, short ears and legs, and a puff-ball tail.

COLOR Tan with black-flecked fur on its body, a rusty orange back of its neck, and a white belly and tail.

BEHAVIOR Mostly crepuscular. It hops in zigzags when it's alarmed to avoid predators, and it stands on its hind legs to look around.

HABITAT Forest edges, old fields, prairies, swamps, neighborhoods, parks, gardens.

POINT OF FACT Unfortunately for them, they're on everyone's menu, such as coyotes, foxes, hawks, owls, bobcats, snakes, weasels, raccoons, and minks. Few Cottontails live more than a year or two in the wild.

I SAW IT!

WHEN I SAW IT

DATE

WHERE I SAW IT

SPECIFIC LOCATION, INCLUDE STATE

WHAT IT WAS DOING

NOTES

....

....

MEDIUM-SIZE MAMMALS

Woodchuck

(Marmota monax)

SIZE 14–20 inches (35½–51 cm) in body length, plus 5–7 inches (12½–18 cm) of tail

SHAPE Large, round, plump rodent with small ears, short legs, and a bushy tail.

COLOR Gray and brown grizzled back, a rusty brown belly, and dark brown legs and tail.

BEHAVIOR It's active in early mornings and late afternoons. It stands up near its burrow and holds its food in its front paws, like a squirrel. It moves on the ground, climbs trees, and shouts a sharp whistle.

HABITAT Fields, roadsides, woodland edges, parks, lawns.

POINT OF FACT You may know it by its other names: the groundhog and whistle pig. It hibernates during the late fall and winter, but wakes up for spring before the snow melts—right around Groundhog Day!

WHEN I SAW IT

DATE

WHERE I SAW IT

SPECIFIC LOCATION, INCLUDE STATE

WHAT IT WAS DOING

NOTES

MEDIUM-SIZE MAMMALS

Yellow-bellied Marmot

(Marmota flaviventris)

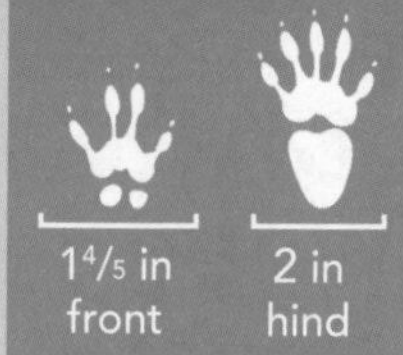

SIZE 13–19 inches (33–48½ cm) in body length, plus 5–8½ inches (12½–21½ cm) of tail

SHAPE Large, stocky rodent with small ears, short legs, and a bushy tail.

COLOR Gray-brown back, a brown head, white on muzzle and over its eyes, a brown tail and feet, and an orange belly and lower legs.

BEHAVIOR Active in early mornings and late afternoons. It hibernates in the winter.

HABITAT Mountain meadows near rock outcrops and rocky slopes. Forages among and sunbathes on rocks.

POINT OF FACT It lives in small family groups that look out for each other, often communicating about dangers with whistles, screams, and teeth-chattering sounds.

I SAW IT!

WHEN I SAW IT
DATE

WHERE I SAW IT
SPECIFIC LOCATION, INCLUDE STATE

WHAT IT WAS DOING

NOTES

MEDIUM-SIZE MAMMALS

Muskrat

(Ondatra zibethicus)

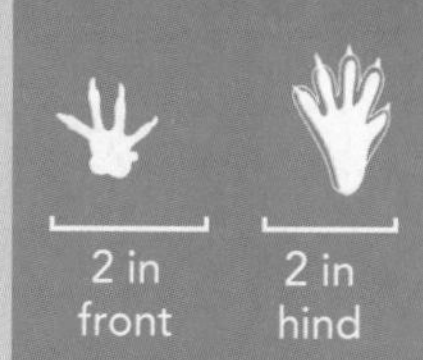

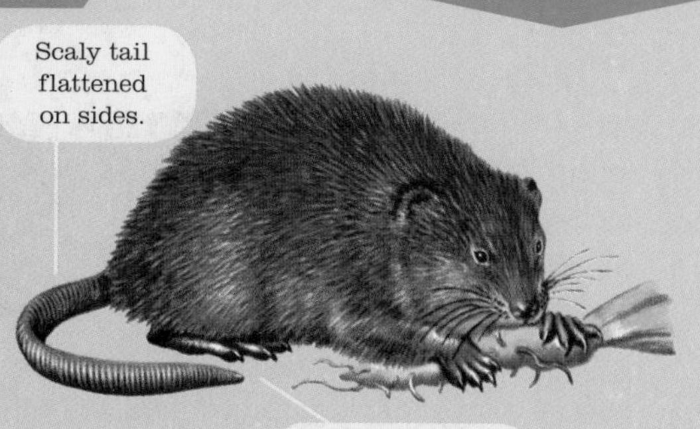

SIZE 10½–14 inches (26½–35½ cm) in body length, plus 7½–10½ inches (19–26½ cm) of tail

SHAPE Medium-size, round, furry rodent with tiny ears, long claws, and a flattened, naked tail.

COLOR Shiny brown fur, a lighter belly, and black eyes.

BEHAVIOR Mostly nocturnal. It's semiaquatic because it spends much of its time in the water to swim.

HABITAT Marshes, streams, ponds.

POINT OF FACT It can close its lips around its large front teeth, allowing it to clip plants without swallowing water.

I SAW IT!

WHEN I SAW IT
DATE

WHERE I SAW IT
SPECIFIC LOCATION, INCLUDE STATE

WHAT IT WAS DOING

NOTES

MEDIUM-SIZE MAMMALS

North American River Otter

(Lontra canadensis)

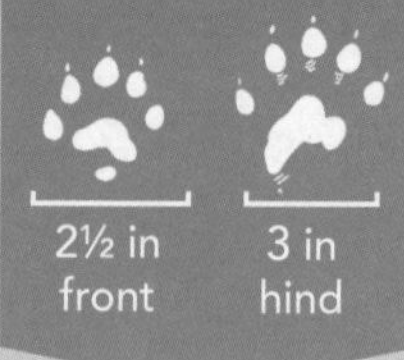

SIZE 26–31 inches (66–78½ cm) in body length, plus 12–20 inches (30½–51 cm) of tail

SHAPE Long, sleek weasel with short legs, small ears, wide feet, and a thick tail.

COLOR Shiny brown back and a silver belly. Its eyeshine is light orange.

BEHAVIOR Crepuscular and semiaquatic. It swims with its head out of the water. It moves over land in a gallop and slides over snow and mud.

HABITAT Streams, rivers, swamps, lakes, and coastal areas.

POINT OF FACT It's an expert swimmer with waterproof fur and flaps of skin that close up its nose and ears.

I SAW IT!

WHEN I SAW IT

DATE

WHERE I SAW IT

SPECIFIC LOCATION, INCLUDE STATE

WHAT IT WAS DOING

NOTES

MEDIUM-SIZE MAMMALS

Long-tailed Weasel

(Mustela frenata)

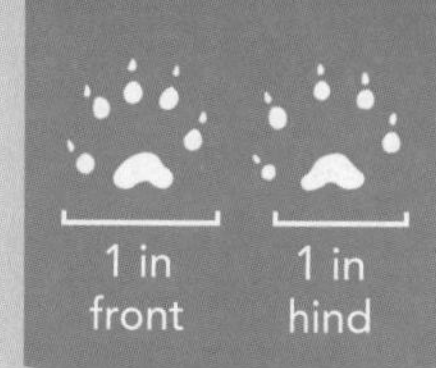

SIZE 7–11 inches (18–28 cm) in body length, plus 4–6 inches (10–15 cm) of tail

Long tail. Tail tipped in black.

SHAPE Large weasel with a long tail and wide, rounded ears.

COLOR Varies by region and season. Usually it's brown on top. In the summer, it has a cream belly to chin; in northern winters it turns white, and in the South it stays brown. Its eyeshine is green.

BEHAVIOR Both diurnal and nocturnal. It gallops across the ground with its back arched and its tail up. It also climbs trees and swims.

HABITAT Woods, fields, meadows, usually near water.

POINT OF FACT In northern areas, it molts dark fur and grows in a white coat, making some good snow camouflage!

I SAW IT!

WHEN I SAW IT

DATE

WHERE I SAW IT

SPECIFIC LOCATION, INCLUDE STATE

WHAT IT WAS DOING

NOTES

MEDIUM-SIZE MAMMALS

American Beaver

(Castor canadensis)

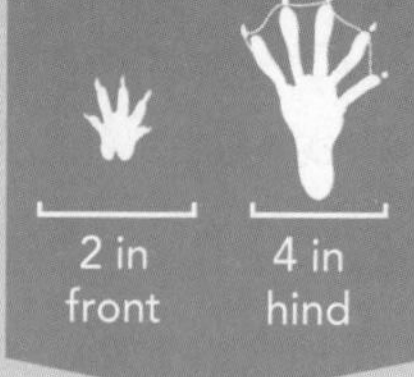

SIZE 24–36 inches (61–91½ cm) in body length, plus 12–15 inches (30½–38 cm) of tail

SHAPE Large, husky, round rodent with small eyes and ears, a large, paddle-shaped tail, and large, webbed hind feet.

COLOR Shiny, red-brown fur, a dark brown tail, and orange front teeth.

BEHAVIOR Nocturnal or crepuscular. Semiaquatic. It swims well and quickly, cuts trees and leaves the stumps behind, makes trails to water, and slaps its tail on the water when it's alarmed.

HABITAT Swamps, lakes, rivers, ponds, and streams in wooded areas.

POINT OF FACT Its front teeth are orange from the iron that strengthens its tree chompers! Wood chips and cone-shaped stumps are signs a beaver has been near.

I SAW IT!

WHEN I SAW IT

DATE

WHERE I SAW IT

SPECIFIC LOCATION, INCLUDE STATE

WHAT IT WAS DOING

NOTES

Nine-banded Armadillo

(Dasypus novemcinctus)

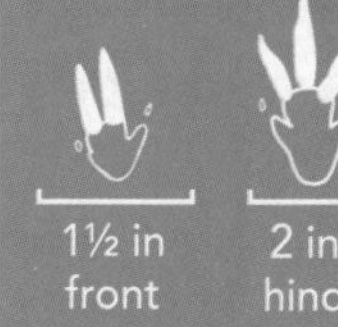

SIZE 15–22 inches (38–56 cm) in body length, plus 10–16 inches (25½–40½ cm) of tail

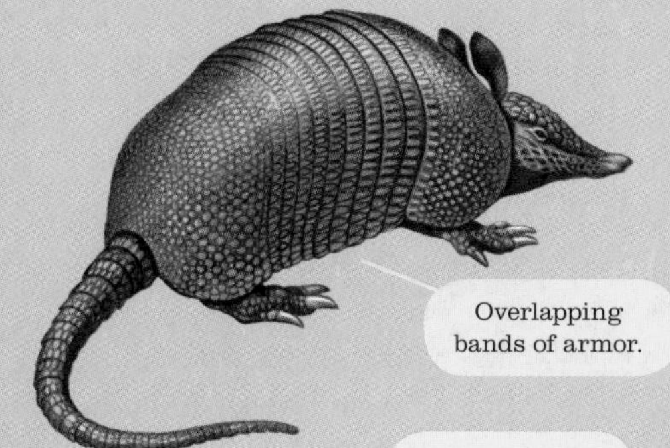

SHAPE Oval shaped with short legs, thick claws, a narrow snout, tall ears, small eyes, and a long, scaly tail.

COLOR Gray-brown scaly covering, a pink nose, and pink toes.

BEHAVIOR Nocturnal in summer, diurnal and nocturnal in winter. It rummages, snuffles, and snorts in leaf litter. It jumps straight up if it's alarmed and runs away with stiff legs.

HABITAT Woodlands, fields, brushy areas, roadsides.

POINT OF FACT It crosses streams by either gulping air to float and swim, or it sinks to the bottom and walks across. It can hold its breath for up to six minutes.

I SAW IT!

WHEN I SAW IT
DATE

WHERE I SAW IT
SPECIFIC LOCATION, INCLUDE STATE

WHAT IT WAS DOING

NOTES

Black-tailed Prairie Dog

(Cynomys ludovicianus)

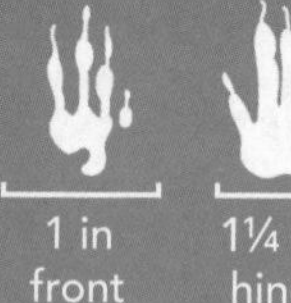

SIZE 11–13 inches (28–33 cm) in body length, plus 2½–3½ inches (6½–9 cm) of tail

SHAPE Beefy prairie dog with medium-size eyes and a long, thin tail.

COLOR Yellow-brown fur with white flecks, light sides and belly, and a black tail tip.

BEHAVIOR Lives in large towns of dug-out burrows and tunnels. It emerges to graze, scan for predators on top of mounds, and socialize. It communicates to others with barks, whistles, and chuckles.

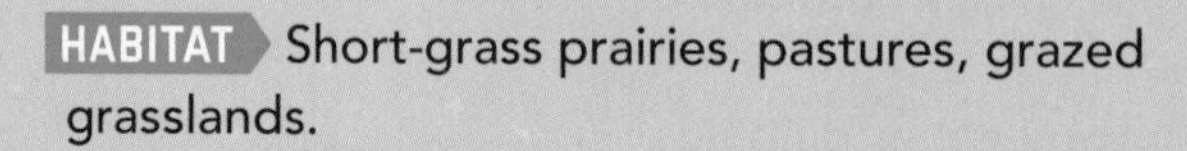

HABITAT Short-grass prairies, pastures, grazed grasslands.

POINT OF FACT Its signature move is the "jump-yip," a territorial call of *weee-ooo* shouted while standing up with its head thrown back.

I SAW IT!

WHEN I SAW IT
DATE

WHERE I SAW IT
SPECIFIC LOCATION, INCLUDE STATE

WHAT IT WAS DOING

NOTES

SMALL MAMMALS

Eastern Gray Squirrel

(Sciurus carolinensis)

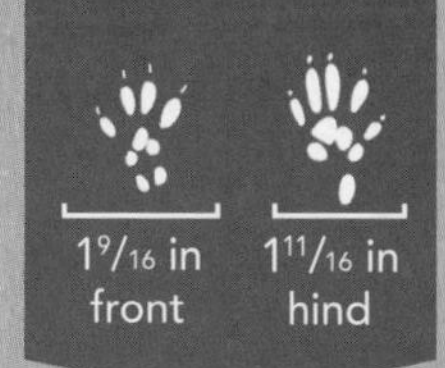

SIZE 9–11 inches (23–28 cm) in body length, plus 6–10 inches (15–25½ cm) of tail

SHAPE Medium-size tree squirrel with a long, bushy tail, medium-size rounded ears, and big, round eyes.

COLOR Gray with some rust-colored areas on its lower sides, face, and tail. It also has a cream belly and chest.

BEHAVIOR Diurnal and solitary. It runs up trees and along branches, jumping acrobatically from tree to tree. It chirrs loudly in alarm and builds leafy nests in trees.

HABITAT Forests, parks, neighborhoods with trees.

POINT OF FACT Its tail has many different uses: It works as a blanket when it's cold, an umbrella in the rain, and a communication tool, too. A fluffed-up tail means it's angry!

I SAW IT!

WHEN I SAW IT
DATE

WHERE I SAW IT
SPECIFIC LOCATION, INCLUDE STATE

WHAT IT WAS DOING

NOTES

SMALL MAMMALS

Eastern Fox Squirrel

(Sciurus niger)

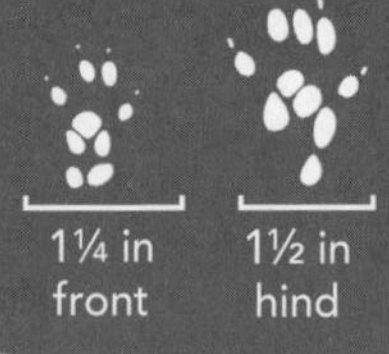

SIZE 10–14½ inches (25½–37 cm) in body length, plus 8–13 inches (20½–33 cm) of tail

SHAPE Large tree squirrel with a long, bushy tail, rounded ears on top of its head, and big, round eyes.

COLOR Orange-brown with a peach-colored belly and chest.

BEHAVIOR Diurnal and solitary. It forages and runs on the ground, and rests and builds leafy nests in trees.

HABITAT Open woodlands, suburbs.

POINT OF FACT It's the largest tree squirrel in the East, weighing up to 3 pounds (1.4 kg).

I SAW IT!

WHEN I SAW IT
DATE

WHERE I SAW IT
SPECIFIC LOCATION, INCLUDE STATE

WHAT IT WAS DOING

NOTES

....

....

SMALL MAMMALS

Red Squirrel

(Tamiasciurus hudsonicus)

1 1/16 in front

1 3/8 in hind

White eye ring.

Brown-orange back.

SIZE 7–8 inches (18–20½ cm) in body length, plus 4–5 inches (10–12½ cm) of tail

SHAPE Medium-size tree squirrel with a medium tail, rounded ears on top of its head, and big, round eyes.

COLOR Dark orange on its back with brown sides, a white belly, and a white eye ring.

BEHAVIOR Diurnal and solitary. It's noisy, and scolds visitors with barks, chirrs, and squeaks as it perches from pine trees.

HABITAT Coniferous forests, mixed forests, parks, orchards.

POINT OF FACT It prefers to eat pine seeds, and it hoards pinecones. Piles of pinecone pieces and nests of shredded bark, lichen, grass, leaves, and twigs are signs it's been around.

I SAW IT!

WHEN I SAW IT

DATE

WHERE I SAW IT

SPECIFIC LOCATION, INCLUDE STATE

WHAT IT WAS DOING

NOTES

....

....

SMALL MAMMALS

Eastern Chipmunk

(Tamias striatus)

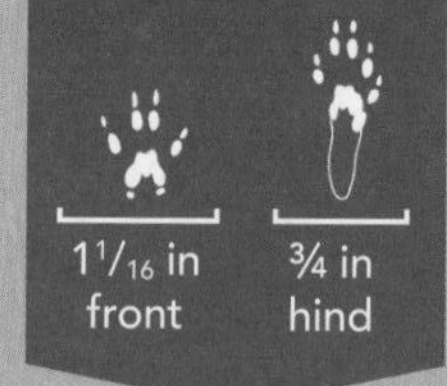

SIZE 4½–6 inches (11½–15 cm) in body length, plus 3–4½ inches (7½–11½ cm) of tail

SHAPE Small ground squirrel with a medium tail, rounded ears on top of its head, large eyes, and a pointed snout.

COLOR Red-brown to gray-brown with a dark stripe down its back, white eye rings, and short, black and white stripes on its sides.

BEHAVIOR Diurnal. It calls out a warning from lookouts on stumps or log piles. It forages on the ground and in trees, and it digs systems of tunnels.

HABITAT Deciduous forest, forest edges, gardens, parks, neighborhoods.

POINT OF FACT It burrows and makes up to 100 feet (30½ m) of underground tunnels with multiple entrances. It also digs short emergency escape burrows.

I SAW IT!

WHEN I SAW IT
DATE

WHERE I SAW IT
SPECIFIC LOCATION, INCLUDE STATE

WHAT IT WAS DOING

NOTES

SMALL MAMMALS

Least Chipmunk

(Neotamias minimus)

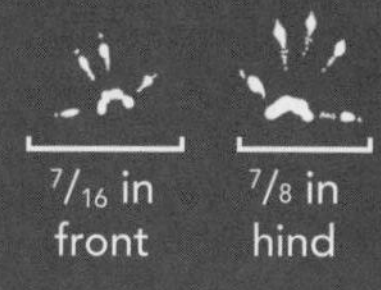

SIZE 4–4½ inches (10–11½ cm) in body length, plus 3–4 inches (7½–10 cm) of tail

SHAPE Small ground squirrel with a long, narrow tail, large rounded ears on top of its head, and big eyes.

COLOR Orange-brown to yellow-gray with dark stripes on its face and down its back.

BEHAVIOR Diurnal. It runs across the ground with its tail held up straight. When stopped, it flicks its tail up and down.

HABITAT Mountain coniferous forests and also sagebrush desert, scrubland, meadows, and alpine tundra.

POINT OF FACT Smallest of all chipmunks, it weighs only 1–2 ounces (28–57 grams), about as much as 5–8 quarters. Great climbers, will scale trees to sunbathe in cool weather.

I SAW IT!

WHEN I SAW IT
DATE

WHERE I SAW IT
SPECIFIC LOCATION, INCLUDE STATE

WHAT IT WAS DOING

NOTES

....

....

SMALL MAMMALS

Yellow-pine Chipmunk

(Neotamias amoenus)

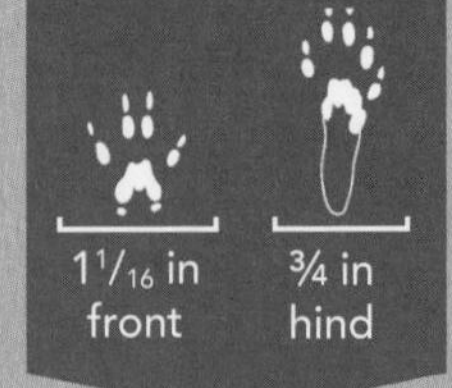

SIZE 4½–5 inches (11½–12½ cm) in body length, plus 3½–4½ inches (9–11½ cm) of tail

Yellow-orange sides.

Tail not as long as body.

SHAPE Small ground squirrel with a narrow tail, rounded ears on top of its head, and big eyes.

COLOR Bright yellow to orange on its shoulders and sides, a cream belly, and a gray or brown rump. It has stripes on its face and sides.

BEHAVIOR Diurnal. Its alarm call is a repeating, high chirp. It runs with its tail out, but not straight up, and it nests in burrows and on tree branches.

HABITAT Open coniferous forest, rocky brush-covered areas.

POINT OF FACT It forages for seeds, fungi, insects, and tubers, and it stores food in underground chambers for winter snacking.

I SAW IT!

WHEN I SAW IT

DATE

WHERE I SAW IT

SPECIFIC LOCATION, INCLUDE STATE

WHAT IT WAS DOING

NOTES

SMALL MAMMALS

Southern Flying Squirrel

(Glaucomys volans)

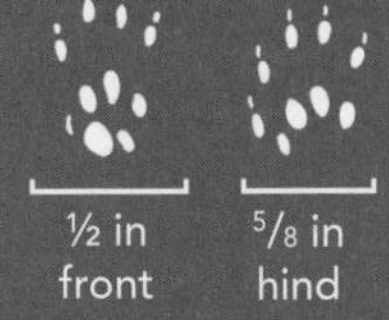

SIZE 4–5½ inches (10–14 cm) in body length, plus 3–4½ inches (7½–11½ cm) of tail

SHAPE Small squirrel with furry skin between its front and hind feet, huge eyes, and a flattened tail.

COLOR Light gray to brown on its back, a white belly, and white cheeks.

BEHAVIOR Nocturnal. It spends most of its time in trees, traveling from one to another by gliding with its legs outstretched.

HABITAT Deciduous forests with acorn-making oaks and nut trees like beech and hickory.

POINT OF FACT Each of its glides averages about 65 feet (20 m), but it can glide up to 300 feet (91½ m) when going downhill. The fur-covered skin between its feet acts like a parachute and its flattened tail like a rudder.

I SAW IT!

WHEN I SAW IT
DATE

WHERE I SAW IT
SPECIFIC LOCATION, INCLUDE STATE

WHAT IT WAS DOING

NOTES

....

....

Golden-mantled Ground Squirrel

(Callospermophilus lateralis)

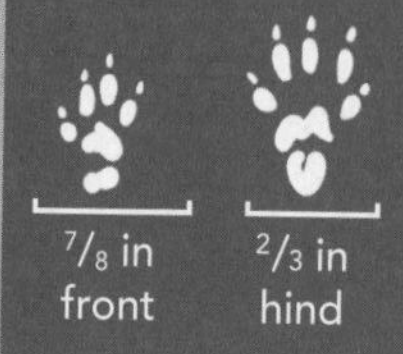

SIZE 6–8½ inches (15–21½ cm) in body length, plus 3–5 inches (7½–12½ cm) of tail

SHAPE Chipmunk-like, small ground squirrel with rounded ears on top of its head, big eyes, a round body, and a medium-length tail.

COLOR Gray-brown back, an orange head, a cream belly, cream lower sides, and double black stripes down its sides with white in between them.

BEHAVIOR Diurnal and solitary. It's often curious and busy, and it perches on tree stumps and picnic tables to eat or stuff seeds in its cheek pouches.

HABITAT Coniferous forests, open mixed woods, alpine meadows, brushy and rocky areas.

POINT OF FACT It sleeps in a burrow it digs, often near the bottom of a tree stump, big rock, or under a fallen log.

I SAW IT!

WHEN I SAW IT
DATE

WHERE I SAW IT
SPECIFIC LOCATION, INCLUDE STATE

WHAT IT WAS DOING

NOTES

....

....

Thirteen-lined Ground Squirrel

(Ictidomys tridecemlineatus)

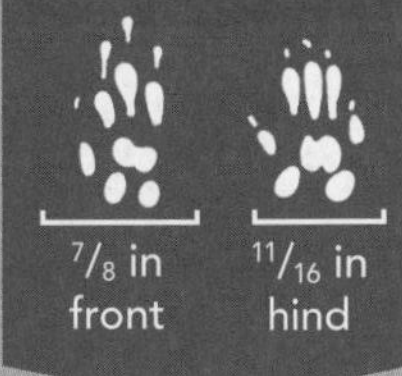

SIZE 6–8 inches (15–20½ cm) in body length, plus 2½–4 inches (6½–10 cm) of tail

SHAPE Slender ground squirrel with small ears on the sides of its head, a long tail, and big eyes.

COLOR Light brown body and feet with rows of dark brown and cream stripes and spots.

BEHAVIOR Diurnal. It builds systems of shallow burrows in open spaces.

HABITAT Grasslands, roadsides, lawns, pastures, short-grass prairie, mountain meadows.

POINT OF FACT It's called a gopher in many places, and during its winter hibernation, its heart rate drops from 200 beats a minute to only 4 or 5.

WHEN I SAW IT

DATE

WHERE I SAW IT

SPECIFIC LOCATION, INCLUDE STATE

WHAT IT WAS DOING

NOTES

....

....

SMALL MAMMALS

Plains Pocket Gopher

(Geomys bursarius)

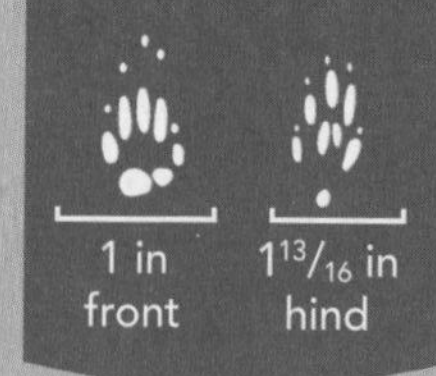

SIZE 6–9 inches (15–23 cm) in body length, plus 2–4½ inches (5–11½ cm) of tail

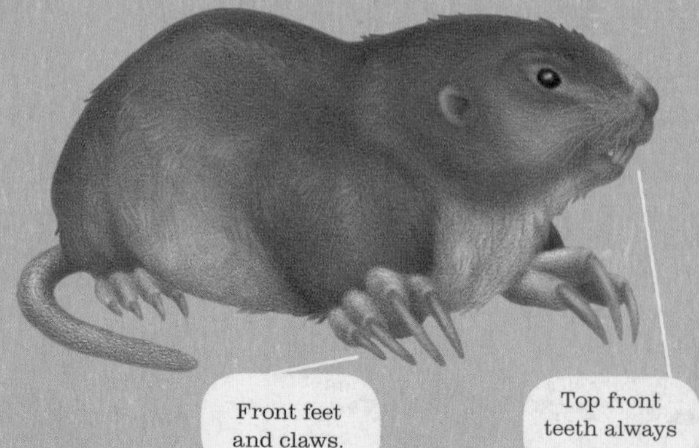

SHAPE Medium-size, round-bodied rodent with tiny ears and eyes, large front teeth, big front legs and claws, and a naked tail.

COLOR Brown on its back and head with a pale belly.

BEHAVIOR Both diurnal and nocturnal. It likes to be solitary and it builds a system of burrows by pushing dirt to the surface before plugging up the hole.

HABITAT Prairies, roadsides, farmlands.

POINT OF FACT It can close its lips behind its front teeth, which keeps dirt out of its mouth while chewing through underground roots.

I SAW IT!

WHEN I SAW IT
DATE

WHERE I SAW IT
SPECIFIC LOCATION, INCLUDE STATE

WHAT IT WAS DOING

NOTES

....

....

SMALL MAMMALS

Northern Short-tailed Shrew

(Blarina brevicauda)

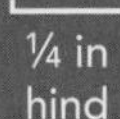

3/16 in front

1/4 in hind

SIZE 3½–4½ inches (9–11½ cm) in body length, plus an inch (2½ cm) of tail

SHAPE Large, plump shrew with small eyes, tiny ears hidden in fur, a pointy snout, and a short tail.

COLOR Silver, dark gray to brown back and head, with a gray belly.

BEHAVIOR Nocturnal and active in the early morning. It digs and tunnels through shallow burrows in the ground, under leaf litter, or under snow.

HABITAT Forests with leaf litter, pond and stream edges, grasslands.

POINT OF FACT Its spit has a poison in it that can paralyze or kill the prey that it bites.

I SAW IT!

WHEN I SAW IT

DATE

WHERE I SAW IT

SPECIFIC LOCATION, INCLUDE STATE

WHAT IT WAS DOING

NOTES

....

....

SMALL MAMMALS

Eastern Mole

(Scalopus aquaticus)

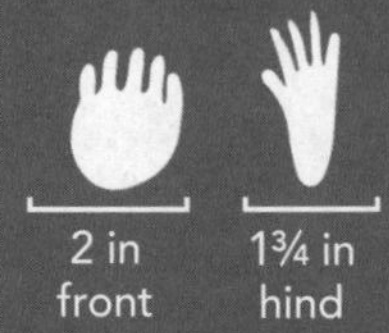

SIZE 4½–6 inches (11–15 cm) in body length, plus 1–1½ inches (2½–4 cm) of tail

SHAPE Wide-bodied, large mole with huge front feet and claws, a long, thin, naked nose, tiny eyes, no ears, and a short, naked tail.

COLOR Gray-brown back and head with a pale belly. It also has a pink, naked nose and tail.

BEHAVIOR Both nocturnal and diurnal. It digs shallow burrows that leave piles of soil (molehills) and ridges on the ground.

HABITAT Woods, lawns, fields with soft, moist dirt.

POINT OF FACT The velvety fur of shrews and moles helps them to move forward and backward underground.

I SAW IT!

WHEN I SAW IT

DATE

WHERE I SAW IT

SPECIFIC LOCATION, INCLUDE STATE

WHAT IT WAS DOING

NOTES

SMALL MAMMALS

Star-nosed Mole

(Condylura cristata)

SIZE 3½–5 inches (9–12½ cm) in body length, plus 2½–3½ inches (6½–9 cm) of tail

SHAPE Wide-bodied, medium-size mole with large front feet, big claws, a long tail, and a nose tipped with a ring of tentacles.

COLOR Dark brown to black body and tail, dark skin, and pink nose tentacles.

BEHAVIOR Both nocturnal and diurnal. It's semiaquatic, so it tunnels to water and swims well, hunting for fish and aquatic invertebrates.

HABITAT Wet woods, swamps, meadows, and near streams.

POINT OF FACT Its tentacles are like super-sensitive fingers that help it search for food in mud and dirt.

I SAW IT!

WHEN I SAW IT
DATE

WHERE I SAW IT
SPECIFIC LOCATION, INCLUDE STATE

WHAT IT WAS DOING

NOTES

....

....

North American Deer Mouse

(Peromyscus maniculatus)

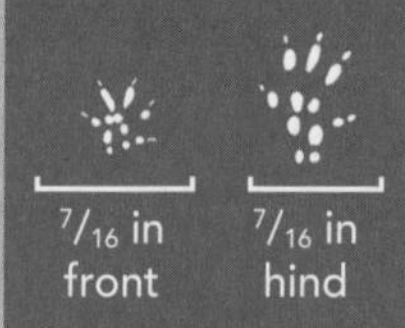

SIZE 3–4 inches (7½–10 cm) in body length, plus 3–4 inches (7½–10 cm) of tail

SHAPE Medium-size mouse with a long, furred tail and largish ears.

COLOR Varies, but it's usually brown (ranging from gray to red) on its back, head, and top of its tail. It has a white belly, chest, chin underside, and bottom of its tail.

BEHAVIOR Nocturnal. It forages for and stores seeds, insects, and other foods.

HABITAT Nearly all, from forests and tundra to swamps, prairies, deserts, and mountains.

POINT OF FACT It builds nests of shredded materials in logs, stumps, sheds, abandoned burrows, old bird nests, rock piles, and anywhere else it can find.

I SAW IT!

WHEN I SAW IT
DATE

WHERE I SAW IT
SPECIFIC LOCATION, INCLUDE STATE

WHAT IT WAS DOING

NOTES

....

....

SMALL MAMMALS

Western Jumping Mouse

(Zapus princeps)

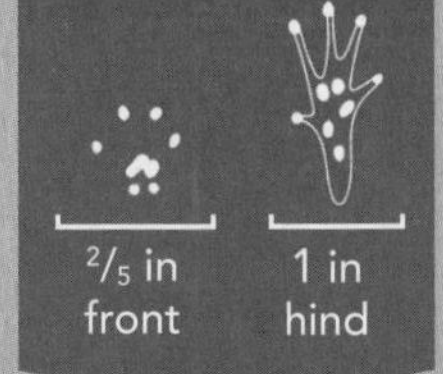

SIZE 3–4 inches (7½–10 cm) in body length, plus 5–6 inches (12½–15 cm) of tail

SHAPE Medium-size mouse with large ears, a very long tail, and long, narrow hind feet.

COLOR Dark brown back with light yellow sides that are flecked with black. It has a white belly, feet, and tail bottom.

BEHAVIOR Mostly nocturnal. It walks on all fours or makes short hops, and it leaps up to 6 feet (2 m) if alarmed.

HABITAT Mountain areas near streams and meadows.

POINT OF FACT It spends 8–10 months hibernating, living off a layer of fat it puts on in autumn.

I SAW IT!

WHEN I SAW IT

DATE

WHERE I SAW IT

SPECIFIC LOCATION, INCLUDE STATE

WHAT IT WAS DOING

NOTES

SMALL MAMMALS

Northern Grasshopper Mouse

(Onychomys leucogaster)

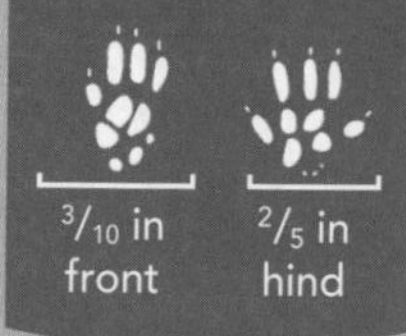

SIZE 4–5 inches (10–12½ cm) in body length, plus 1–2 inches (2½–5 cm) of tail

SHAPE Large mouse with a short tail, big, round eyes, and rounded ears on top of its head.

COLOR Varies from gray to golden brown on its back and head. It also has a white belly, chest, throat, and chin. Its tail is brown on top and white below.

BEHAVIOR Nocturnal and solitary. It stands on its hind legs and makes a loud, one-note whistle to announce its territory.

HABITAT Deserts, grasslands, scrubland, prairies.

POINT OF FACT It hunts grasshoppers and other large insects, scorpions, small lizards, and other mice. It also eats seeds and plants.

I SAW IT!

WHEN I SAW IT

DATE

WHERE I SAW IT

SPECIFIC LOCATION, INCLUDE STATE

WHAT IT WAS DOING

NOTES

....

....

SMALL MAMMALS

Meadow Vole

(Microtus pennsylvanicus)

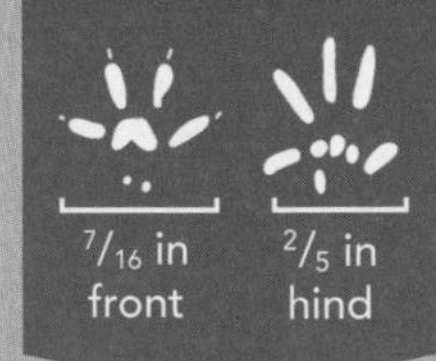

SIZE 4–5½ inches (10–14 cm) in body length, plus 1–2 inches (2½–5 cm) of tail

SHAPE Small but plump rodent with a large head, a short tail, short legs, and rounded ears on the sides of its head.

COLOR Dark brown back, gray-brown sides, and a cream belly.

BEHAVIOR Nocturnal and diurnal. It swims and creates networks of runways through grass, leaving piles of clippings behind.

HABITAT Grassy areas like meadows, roadsides, fields, orchards.

POINT OF FACT A female can give birth to a litter of pups every three weeks, and her offspring are grown adults themselves and ready to breed at one month old. This means populations can explode with enough food and space.

I SAW IT!

WHEN I SAW IT
DATE

WHERE I SAW IT
SPECIFIC LOCATION, INCLUDE STATE

WHAT IT WAS DOING

NOTES

SMALL MAMMALS

House Mouse

(Mus musculus)

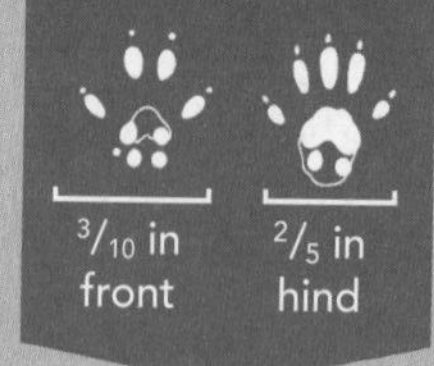

SIZE 3–4½ inches (7½–11½ cm) in body length, plus 2½–3½ inches (7½–9 cm) of tail

No sharp line of color change between belly and sides.

SHAPE Medium-size mouse with a small head, large, naked ears, big, round eyes, and a long, fairly thick tail with sparse fur.

COLOR A brown-gray back and head, a pale belly, and a brown tail.

Tail not white on bottom.

BEHAVIOR Nocturnal. It lives in colonies of several females, a male, and their young. It marks its territory with a musky urine.

HABITAT Farmlands, roadsides, buildings in cities and neighborhoods.

POINT OF FACT It's native to Asia, but it has spread worldwide. It arrived in North America with European explorers and settlers.

I SAW IT!

WHEN I SAW IT
DATE

WHERE I SAW IT
SPECIFIC LOCATION, INCLUDE STATE

WHAT IT WAS DOING

NOTES

....

....

SMALL MAMMALS

Norway Rat

(Rattus norvegicus)

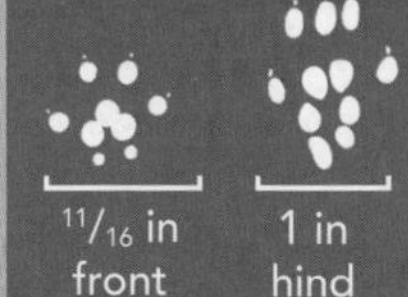

SIZE 7–10 inches (18–25½ cm) in body length, plus 6–8½ inches (15–21½ cm) of tail

SHAPE Large, stocky rat with small ears and a medium-length, sparsely furred tail.

COLOR Yellow-brown back and head, a pale belly, and a brown, nearly naked tail.

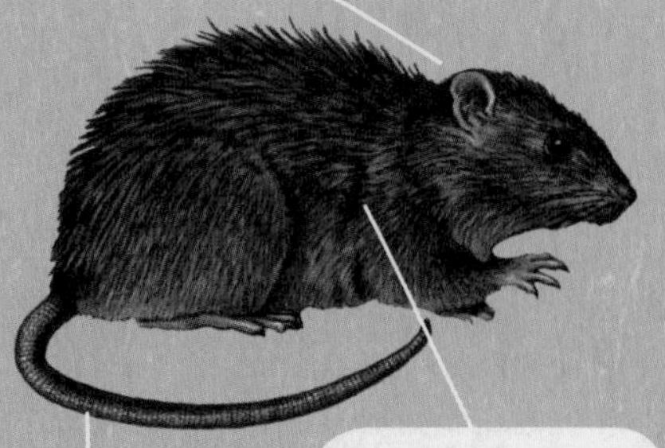

BEHAVIOR Nocturnal. It usually runs on the ground, but it can climb and it's an expert swimmer. Colonies live in systems of burrows.

HABITAT Towns, cities with sewers, grain fields, marshes.

POINT OF FACT It's the most destructive pest mammal because it eats stored grain and food, chews up building structures, and spreads diseases like the plague.

I SAW IT!

WHEN I SAW IT

DATE

WHERE I SAW IT

SPECIFIC LOCATION, INCLUDE STATE

WHAT IT WAS DOING

NOTES

SMALL MAMMALS

Eastern Woodrat

(Neotoma floridana)

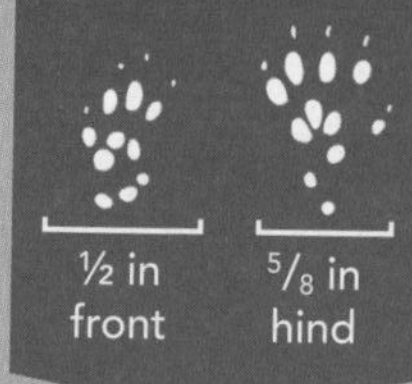

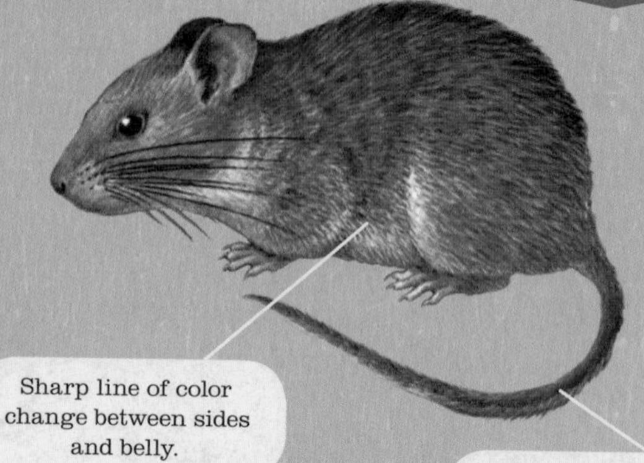

SIZE 7–11 inches (18–28 cm) in body length, plus 5–7 inches (12½–18 cm) of tail

SHAPE Large, plump rat with large, rounded ears, big, round eyes, thick, long whiskers, and a furred tail.

COLOR Gray to sandy brown back and head, lighter sides, belly, and tops of its feet, white tail bottom, and brown tail top.

BEHAVIOR Nocturnal and solitary. It runs on the ground and climbs up into trees. It drums its hind feet when alarmed.

HABITAT Woodlands, rocky areas, swamps.

POINT OF FACT It's also called a "pack rat." It makes a large house of sticks and other collected materials, sometimes just for decoration.

I SAW IT!

WHEN I SAW IT

DATE

WHERE I SAW IT

SPECIFIC LOCATION, INCLUDE STATE

WHAT IT WAS DOING

NOTES

SMALL MAMMALS

Ord's Kangaroo Rat

(Dipodomys ordii)

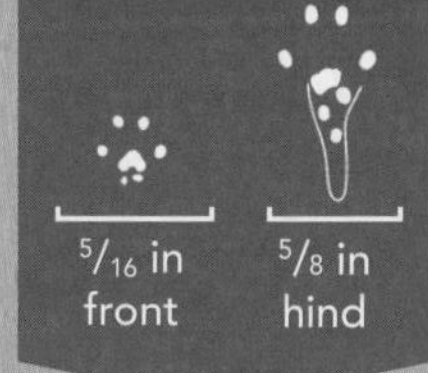

SIZE 3½–4½ inches (9–11½ cm) in body length, plus 4–6 inches (10–15 cm) of tail

SHAPE Small rat with a compact body, an oversize head, huge, round eyes, small ears on the sides of its head, big back feet, and an extremely long tail.

COLOR Golden brown to gray back and head, pale sides, and a white belly. It also has white spots above eyes and a dark stripe along its tail.

BEHAVIOR Nocturnal and solitary. It walks on all fours and hops on its back legs like a kangaroo, then bathes in sand to clean its fur. It drums its hind feet to communicate.

HABITAT Desert scrub, sagebrush, dry grasslands, dry coniferous woods.

POINT OF FACT It's a desert survivor that can live on just the water it gets from food. It eats seeds, some plants, and occasionally insects.

WHEN I SAW IT

DATE

WHERE I SAW IT

SPECIFIC LOCATION, INCLUDE STATE

WHAT IT WAS DOING

NOTES

....

....

Eastern Red Bat

(Lasiurus borealis)

Reddish fur.

Furry tail membrane.

SIZE 3¾–5 inches (9½–13 cm) in body length, plus 1¾–2½ inches (4½–6 cm) of tail

SHAPE Medium bat with long pointed wings, short rounded ears, and a furred tail membrane.

COLOR Bright orange to reddish-brown fur with white patches on shoulders and thumbs. Females are often duller red.

BEHAVIOR Nocturnal and crepuscular. It is a solitary bat that roosts in trees and often begins hunting insects on the wing before total darkness.

HABITAT Fields, woodlands, towns.

POINT OF FACT It commonly hangs by a single foot when roosting in trees during the daytime. It wraps its furry tail membrane around itself and looks like a dead leaf.

WHEN I SAW IT

DATE

WHERE I SAW IT

SPECIFIC LOCATION, INCLUDE STATE

WHAT IT WAS DOING

NOTES

SMALL MAMMALS

Big Brown Bat

(Eptesicus fuscus)

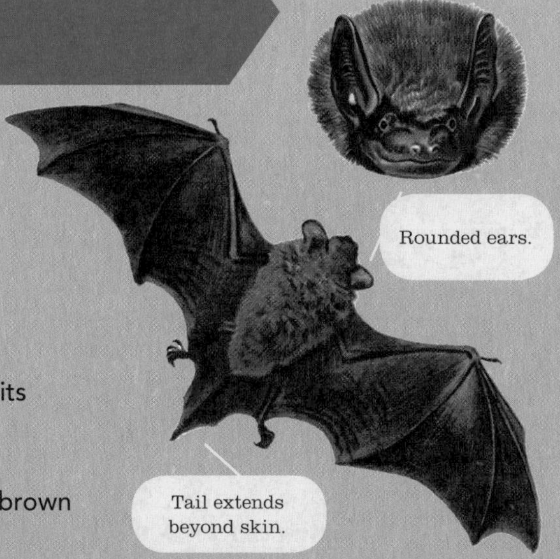

SIZE 2½–3 inches (6½–7½ cm) in body length, plus 1½–2 inches (4–5 cm) of tail

SHAPE Large bat with long fur, medium-size ears with rounded tips, a wide muzzle, and a tail that sticks out beyond the skin between its legs.

COLOR Light brown fur with a dark brown face, ears, and skin.

BEHAVIOR Nocturnal. It comes out after sunset and catches beetles and other flying insects while rapidly beating its wings.

HABITAT Woodlands, fields, neighborhoods.

POINT OF FACT It roosts in barns, attics, and hollow trees. It hibernates in caves, mines, or attics.

I SAW IT!

WHEN I SAW IT

DATE

WHERE I SAW IT

SPECIFIC LOCATION, INCLUDE STATE

WHAT IT WAS DOING

NOTES

....

....

Mexican Free-tailed Bat

(Tadarida brasiliensis)

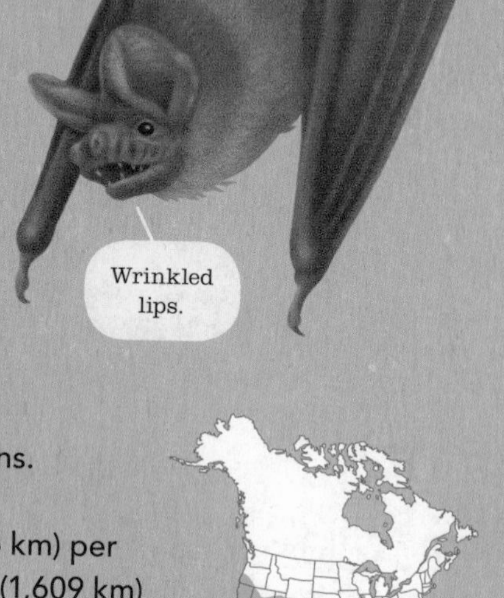

SIZE 2–2½ inches (5–6½ cm) in body length, plus 1–1½ inches (2½–4 cm) of tail

SHAPE Large bat with big ears that meet above its forehead, wrinkled lips, long, narrow wings, and a tail that sticks out past the skin between its legs.

COLOR Brown fur and skin that's paler on its belly.

BEHAVIOR Nocturnal. It lives in groups that roost in caves, under bridges, and in buildings. It hunts moths and other flying insects.

HABITAT Desert, scrubland, farmland, towns.

POINT OF FACT It can fly up to 60 miles (96 km) per hour and migrates as many as 1,000 miles (1,609 km) to its winter home in Mexico.

WHEN I SAW IT

DATE

WHERE I SAW IT

SPECIFIC LOCATION, INCLUDE STATE

WHAT IT WAS DOING

NOTES

....

....

SMALL MAMMALS

Hoary Bat

(Lasiurus cinereus)

SIZE 3–3½ inches (7½–9 cm) in body length, plus 2–2½ inches (5–6½ cm) of tail

SHAPE Large bat with long fur, medium-size, rounded ears, and a furred tail tip within the skin between its legs.

COLOR Body fur has dark roots with frosted tips. It has a black nose and ear edges, a yellow ruff of fur around its face, and white fur patches near its thumbs and shoulders. Young are grayish with frosted tips.

BEHAVIOR Nocturnal. It catches moths and other large flying insects over streams and ponds and around streetlights.

HABITAT Deciduous and coniferous forests, usually near water.

POINT OF FACT Its powerful wings allow it to migrate long distances. It's the only native Hawaiian mammal, the only furry creature to have reached the islands on its own. All the others came with humans.

WHEN I SAW IT

DATE

WHERE I SAW IT

SPECIFIC LOCATION, INCLUDE STATE

WHAT IT WAS DOING

NOTES

PART IV

AMPHIBIANS AND REPTILES

You're thinking about supper as you hike back toward the trailhead. It's starting to get dark—and chilly. You stop to dig out a flashlight and sweatshirt, setting your backpack down next to a big rock. Might as well have some water, too. Thirst quenched, sweatshirt on, you reach down for the backpack. That's when you see the snake next to the same big rock. A snake with red, black, and yellow bands. Oh, boy. Is it venomous? Coral snakes are red, black, and yellow—and they're venomous. But king snakes are red, black, and yellow, too. And king snakes are harmless! Is it safe to grab your backpack? Would you be able to tell the two apart? *What would you do?*

CHAPTER 1

How to Find Herps

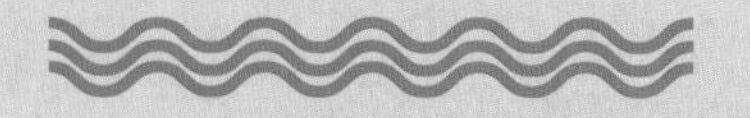

Searching for frogs, snakes, toads, and turtles means looking in places you might usually ignore. Like where? Around piles of rocks and under logs, or in weedy ponds and muddy ditches. Why? Because reptiles and amphibians lead lives very different from yours. Imagine having a baby sister that changes from an aquatic creature into a land-living one, like a frog does! Think about eating only a few times a month, like snakes do, and only being active in warm temperatures. So much for winter sports! And most reptiles and amphibians hatch from eggs without a parent in sight, surviving on their own (or not!) from birth. That's the life of most herps, another name for amphibians and reptiles.

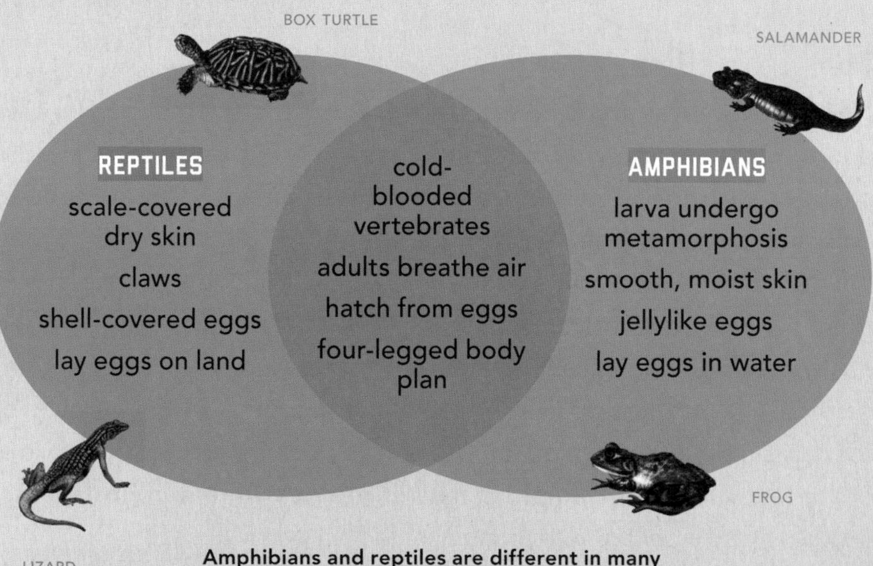

Amphibians and reptiles are different in many ways, but both are cold-blooded vertebrates.

CRITTER CONFUSION: Lizard or Salamander?

Salamanders and lizards are both four-legged, cold-blooded animals with tails. Here's some clues to tell the difference.

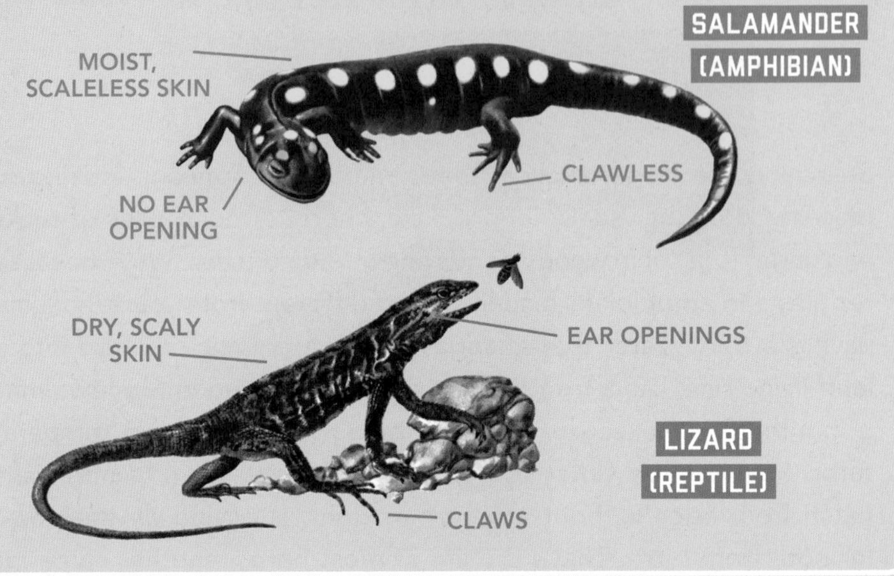

Herp Basics

Herpetology is the study of reptiles and amphibians, or herps. Reptiles and amphibians are cold-blooded animals. Their body temperature changes with the temperature of the environment. (Warm-blooded animals maintain a steady body temperature independent of their environment.) Being cold-blooded makes living in low temperatures (think: wintertime or the Arctic) difficult. But being warm-blooded has a cost. Fueling a constant body temperature requires constant eating. A warm-blooded rat or robin can't survive on the amount of food a similarly sized frog or turtle does. Being cold-blooded has advantages! After all, amphibians and reptiles were around 100 million plus years before mammals evolved on earth.

One of the first animals to walk on land about 370 million years ago was an amphibian-like creature called Ichthyostega, as shown in this illustration from 1962.

From Tadpole to Frog

Let's start with the slimy stuff! Amphibians include frogs, toads, salamanders, and newts. Amphibians are amphibious (duh!), meaning they inhabit both water and land. If you want to spot amphibians, be prepared to get wet—or at least a bit muddy. Most amphibians start off life in a watery place, hatching from jellylike eggs that need to stay wet. An amphibian's first form is usually aquatic, like a tadpole, complete with gills and a finned tail for swimming. This larval form changes into an adult during metamorphosis. Tadpoles or larval salamanders sprout legs, develop lungs, absorb tails, and become adults.

TIGER SALAMANDER METAMORPHOSIS

Some kinds of terrestrial salamanders lay eggs in moist soil. These go through their larval stage while still inside the egg, hatching out as mini versions of their parents.

Damp Dwellers

Adult amphibians need a moist environment, too. They breathe with lungs, but also take in oxygen (and soak up drinking water) through the smooth, scaleless skin covering their bodies. Survival depends on not drying out, and a nice slimy layer of mucus helps with that. Frogs and other amphibians are much more common in wet or humid places than in deserts and arid regions. There's a reason the damp, foggy Smoky Mountains have more salamander species than anywhere else.

Amphibians generally see and hear well. Most adults are carnivores, feeding on anything that they can catch and swallow. Aquatic tadpoles are more omnivorous, munching on plants and algae as well as whatever else they can gobble—including siblings! Toads, salamanders, and frogs are in turn perfectly snackable prey for many kinds of animals from birds to raccoons.

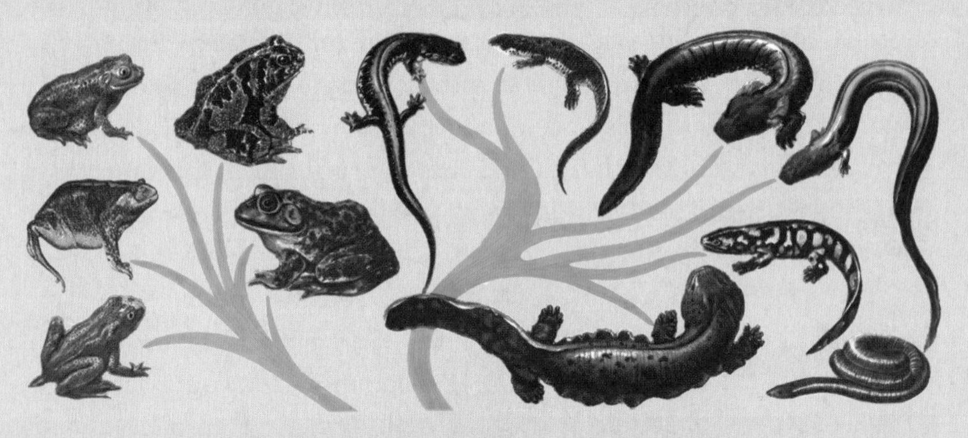

The United States and Canada combined are home to about 300 amphibian species.

Warm Up to Reptiles

Snakes, lizards, turtles, and other reptiles are better equipped for life on dry land than amphibians. Lizard spotting is a sunny day activity! Warm climates suit reptiles, and many species thrive in deserts. The tough scales that cover reptile bodies seal in moisture and act as protective armor. Toes with claws and jaws with ripping teeth equip reptiles while hunting and eating large prey. In fact, a Nile crocodile can clamp on to a zebra and hold it underwater until it drowns.

The United States and Canada have about 430 reptile species, while three times as many make warmer Mexico and Central America their home.

Because their body temperature changes with the environment, reptiles sunbathe or bask—which is a great time to spot them. Soaking up heat from sunshine or a warm rock powers up a reptile so it can hunt, digest food, or find a mate. To cool down, a reptile seeks shade or heads underground. Where winters are harsh, reptiles (amphibians, too!) hibernate until the warmth of spring returns.

Red-eared sliders basking.

Eating and Egg Laying

Most reptiles are predators, eating whatever animals they can catch. However, some turtles and lizards also eat plants. Being cold-blooded means only needing a big meal once or twice a month for many reptiles. (Going to where herps find their food isn't as surefire a way to find them as it is birds or mammals.) A snake takes a long time to digest a swallowed rat. Also, reptiles never stop growing. Their bones and body continue to enlarge as they age, if well fed. New scales constantly form under the old, shedding the outer skin to allow renewal and growth.

Copperhead shedding

The eggs of reptiles are laid on land. The leathery shells of reptile eggs hold in water and protect them. The eggs depend on heat from the sun or rotting plants to incubate, not the body heat of the parent, as with birds. Few reptiles protect their eggs or care for their young, alligator mothers being a dedicated exception (page 381). A few kinds of snakes give birth to live young.

All turtles lay their eggs on land.

Happy Herp Hunting

In general, reptiles like sunny, dry places (left), while amphibians go for dark or shady damp-to-wet spots. There are exceptions (water snakes, desert toads), but if you're lizard hunting, a pond isn't a great place to start.

Reptiles and amphibians are generally shy, secretive, and wary of people. But that only ups the excitement when a snake, frog, turtle, lizard, or salamander is spotted! Successfully stalking and seeing herps takes knowledge of their regular haunts as well as an understanding of their habits and habitats. Here are some herp-finding strategies:

BE WEATHER AWARE Herps and cold temperatures don't mix. Look for them during weather warm enough for them to come out of hiding and heat their bodies.

SEEK SUNNY HOT SPOTS When reptile searching, think about where the sun is. Basking reptiles will more likely be on a sun-lit side of a fence than the shady side. Same for rock piles out in the open rather than rocks shaded by trees. When and where it's superhot (think: Death Valley), mid-morning and late afternoon are prime times as many reptiles head for shade during the scorching midday.

EASTERN FENCE LIZARD

STOP, LOOK, AND LISTEN Stop often and look around with binoculars for sunbathing reptiles. Listen for scampering lizards. Look for moving grass. Look and listen for lizards skittering up pine trees, around rocky cliffs, stone walls, and fence posts.

TRACK TURTLES Find water turtles (and water snakes) basking on logs and rocks in creeks and lakes. Softshell turtles sunbathe on sandy or muddy banks. Box turtles settle into cool muddy puddles on hot days.

PAINTED TURTLES

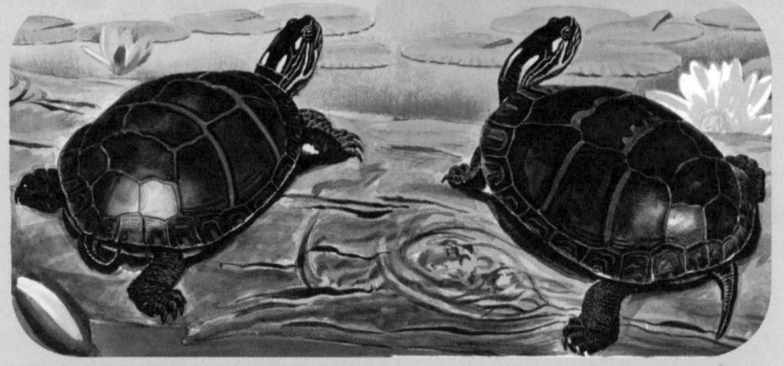

ROLL LOGS AND TILT UP ROCKS Salamanders and newts (and some snakes) hang out in the moist soil under rocks and logs. It's always a good idea to roll a log toward you to give critters a chance to escape in the opposite direction of your feet. Gently put moved logs or rocks back as you found them.

SEARCH IN THE SPRING

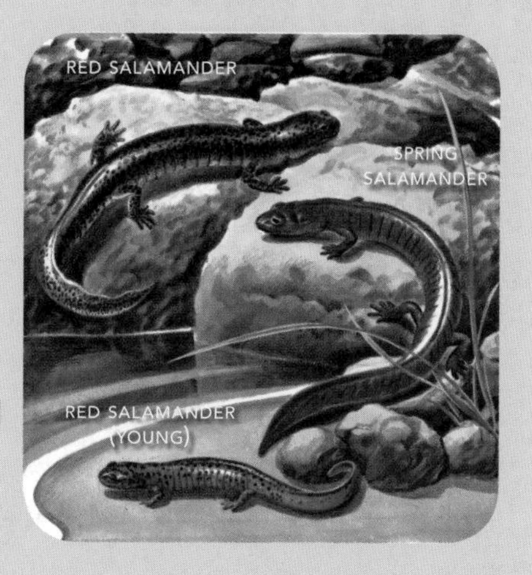

Amphibian hunting is easiest during breeding season when frogs, salamanders, and toads gather to mate and lay eggs. Look for them on the edges of streams and ponds, in wet meadows, and where rain makes temporary pools. Check both the water and the surrounding plants and land. Listen for calling frogs and toads.

WATCH AT NIGHT

Many frogs and salamanders are nocturnal. Avoiding sunshine helps them stay moist. Strap on a headlamp and look for amphibians in the dark under rocks and logs, near streams, in wet fields, and breeding pools. A rainy night is even better!

TRY IT → Herp Habitat Survey

Scientists explore and study a place by doing a survey of it. They record the place's features and write down their findings. You can survey a nearby area to build herp habitat identification skills. Ready?

WHAT YOU'LL NEED

- a pencil or pen

STEP 1 Go to a park or other large natural area that has a variety of habitats—variety is more important than size. In fact, a backyard with a rocky area and a pond is better than a park that's all lawn. Get it?

STEP 2 Walk around the area. Check off what you see:

- ☐ open lawn
- ☐ pond
- ☐ brush or log piles
- ☐ single trees
- ☐ creek
- ☐ sunny rocks
- ☐ mossy areas
- ☐ muddy spots
- ☐ stone walls
- ☐ marsh
- ☐ damp ditches
- ☐ woods or grove of trees
- ☐ other water ....

STEP 3 As you survey the conditions think about where different kinds of reptiles (lizards, snakes, turtles) and amphibians might live. (See Happy Herp Hunting on page 311.) Think about where it's shady or sunny, dry or wet.

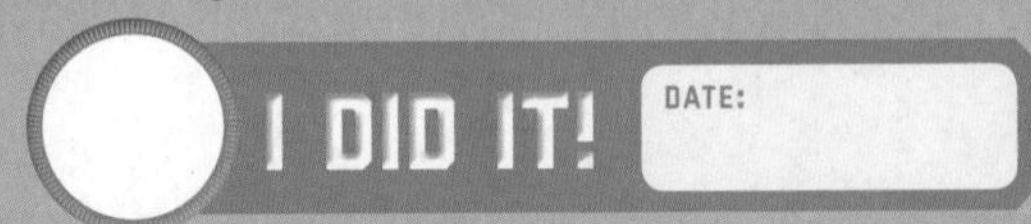

CHAPTER 2

How to ID Frogs, Toads, and Salamanders

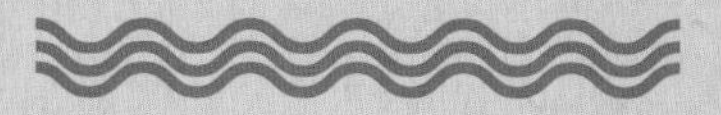

Now that you've learned where to look for frogs, toads, and salamanders, you'll want to identify those you see. Just as with other wildlife, identifying amphibians is about putting these five clues to use.

Here's what to pay attention to:

SHAPE Frogs and salamanders are the two main groups of North American amphibians. Their biggest difference in shape is tail or no tail So if the adult has a tail? Salamander. No tail? Frog. (It's an easy test, folks.)

SIZE Amphibians don't vary as much in size as mammals (shrew to whale!). But bullfrogs are big frogs—you need two hands to hold one—while spring peepers are as small as a thumb, and many terrestrial salamanders aren't much wider than a caterpillar.

SPRING PEEPER

The eastern red-backed salamander comes in both red and black color phases, called morphs.

COLOR & COLOR PATTERNS Unlike bird feathers or fur, the color of frog or salamander skin can vary and change! A single amphibian species often comes in a range of hues. The difference between belly or underside color and its topside or back color is important, too. Two salamanders might look identical until you see their bellies.

Because color isn't always a reliable clue, color patterns and field marks like stripes, spots, lines, and patterns are essential for identification. Frogs can have telltale ridges, skin folds, or eardrums. Some salamanders have specific numbers of rib grooves on their sides, like tiny spare tires.

BEHAVIOR A salamander doesn't take a flying leap into a pond. Nor does a toad leave a tail-drag track in sand. Knowing who hops and who shuffles helps identify amphibians, as does learning their preferred hiding spots and daily routine. Communication counts, too, especially with frogs. (Most salamanders are silent.) Croaks, *ribbits*, and other calls are crucial for identifying frogs, just as with birds.

RANGE & HABITAT Use a field guide with range maps! Even if a salamander looks a lot like the field guide picture, it's probably not that particular critter if where you saw it doesn't match the map.

CRITTER CONFUSION: Frog or Toad?

All toads are frogs, but not all frogs are toads. And not all frogs (like tree frogs) are members of the "true" frog family. Confused? Here are some clues to tell the difference.

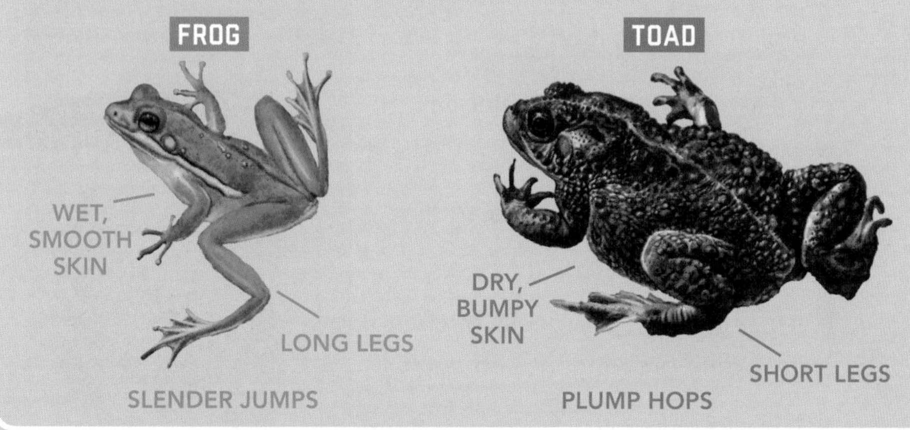

Talk Like an Amphibian Expert

Here are some terms to know when talking toads and salamanders.

ANURAN scientific group for frogs, toads, spadefoots, tree frogs, or other tail-less amphibians

AQUATIC living in water

COSTAL GROOVES rib ridges on sides of some salamanders

DIURNAL active during day

DORSOLATERAL FOLD paired ridge down back of some frogs

ECTOTHERMIC cold-blooded

EFT juvenile stage of a newt after larva

FIELD MARK distinctive feature useful in identification

GILLS organs for breathing in water

LARVA immature amphibian that hatches from egg

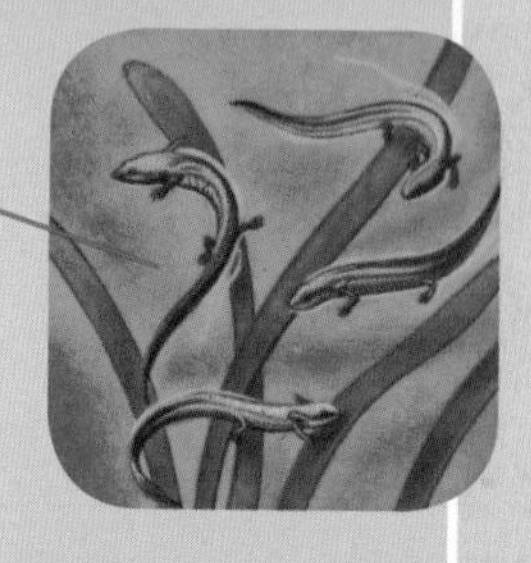

METAMORPHOSIS process of changing forms while growing into an adult

NEWT kind of aquatic salamander with terrestrial juveniles

NOCTURNAL active at night

SALAMANDER scientific group that includes newts, mudpuppies, and sirens

SEMIAQUATIC partially living in water

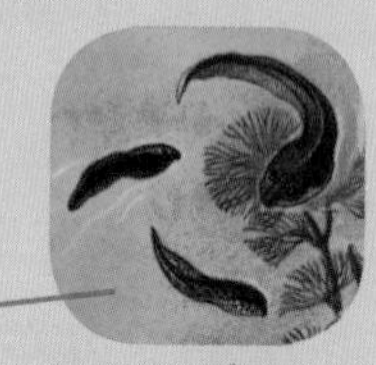

TADPOLE larval aquatic frog or toad

TERRESTRIAL living on land

TOEPAD flesh at end of toe

TYMPANUM eardrum

VOCAL SAC male frog's noise-making, expanding throat pouch

TRY IT → Catch and ID an Amphibian

Frogs (or toads) are probably the easiest wild amphibian to catch and observe. But if you know of a pond full of newts, go for it! (See Happy Herp Hunting tips on page 311.)

WHAT YOU'LL NEED

- a clear jar

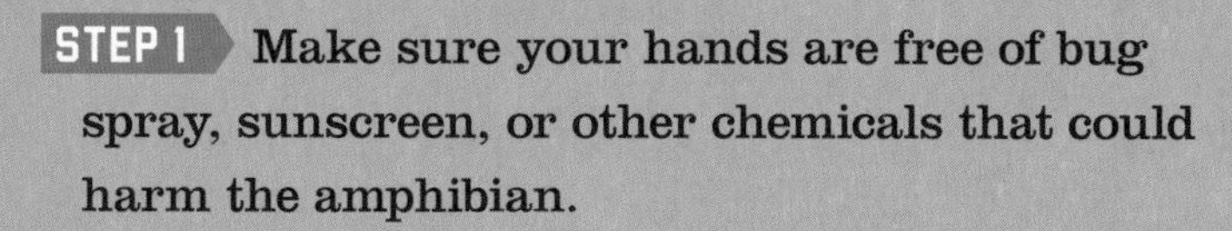

STEP 1 Make sure your hands are free of bug spray, sunscreen, or other chemicals that could harm the amphibian.

STEP 2 Be gentle! Try to scoop it up with your hands, not grab at it. Better that it get away than get squished.

STEP 3 Set it carefully in a jar to look at it, but don't leave it in there more than ten minutes. You don't want it to dry out.

STEP 4 Notice its overall shape, size, and color. Does it have spots or lines, stripes or blotches?

STEP 5 Decide what its *most* distinctive characteristics—its field marks—are. Sloping shape of its back? Huge toepads? Bright red belly?

STEP 6 Once you are done observing the animal, return it where you caught it. And do it in a timely manner.

STEP 7 Wash your hands thoroughly. The skin of some amphibians carries toxins and/or infectious germs like salmonella.

I DID IT! DATE:

TRACK IT ↘ Tracking Amphibians

Now that you've gotten to know an amphibian up close and personal, record what you observed.

NORTHERN LEOPARD FROG

WHAT YOU'LL NEED

- a pencil or pen

DATE

TIME

LOCATION

WEATHER

Describe habitat

Group ☐ frog ☐ toad ☐ salamander

Estimate length of body ....

and tail (if salamander) ....

Do you think it's an adult? ☐ yes ☐ no ☐ not sure

Color

Stripes, spots, or lines?

Any field marks?

What's it doing?

How does it feel?

What sounds does it make?

Solitary or in a group?

Describe habitat

Draw the amphibian and label its field marks.

Nice work!

Can you identify the amphibian? Look for it in the Amphibian Identification section (pages 338–357). Make a positive ID? Check off its I SAW IT! box and fill in the blanks. Congratulations!

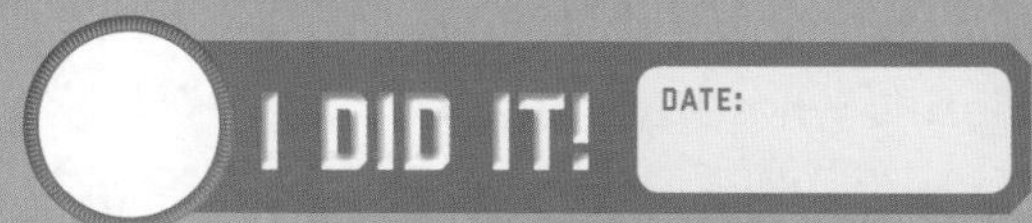

TAKE IT TO THE NEXT LEVEL ↗

Night Watch

Many amphibians stay hidden under rotten logs and damp leaf litter during the day. Sunshine saps their skin of moisture, so being nocturnal helps them survive. Plus, nighttime is bug time—or dinnertime for eaters of insects like frogs and salamanders. After dark is also a great time for hunting amphibians. Especially a rainy spring night when they come out to breed. Strap on a headlamp and look for salamanders under rocks and logs, and frogs and toads near streams, in wet fields, and breeding pools. Make sure to bring an extra flashlight and batteries and dress to get wet and muddy. It's part of the fun!

CHAPTER 3

How to ID Snakes, Lizards, and Turtles

Knowing where to look for snakes, lizards, and turtles is only step one. Being able to identify a garter snake or fence lizard is what's next! Recognizing reptile species is about paying attention to these five clues.

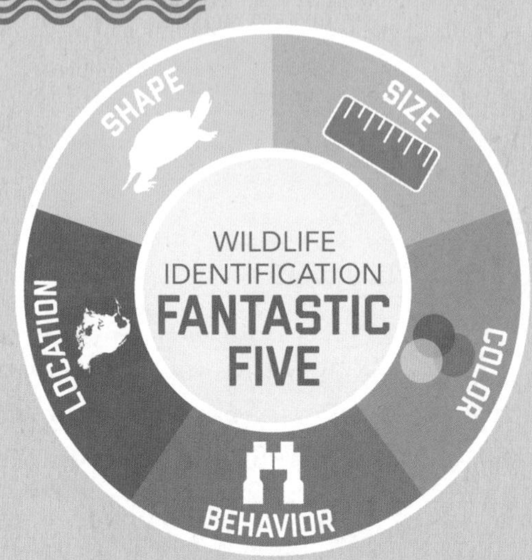

Here's what to focus on:

GROUP SHAPES While they come in dozens of different species, you know a snake when you see one. Same goes for turtles, lizards, and alligators—the other North American reptile groups in the Reptile Identification section (pages 358–381). Note that tortoises, terrapins, and sliders are all kinds of turtles.

SIZE When taking length into consideration, keep in mind that most young reptiles are simply miniature versions of their parents. And also that reptiles are indeterminate grow-ers, which means they never totally stop growing. A six-foot

AMERICAN ALLIGATOR

(1.8 m) long alligator is an adult. But it'll keep growing if there's plenty to eat. This is why reptile sizes in the Reptile Identification section (pages 358–381) vary, and why record sizes (19 feet [6 m] for American alligator) are even bigger! Relative sizes work for reptiles, too. Good lengths to measure and know are the length of your arm, span of your hand, height of your knee—you're a walking yardstick!

EASTERN CORAL SNAKE

COLOR & COLOR PATTERNS Reptile scales come in distinctive colors and patterns often used for identification, even though color can vary quite a bit within a species. The shape of a reptile's scales—keeled, smooth, bead-like—doesn't vary within species. All skinks have smooth scales, for example, while alligators have bumpy scales.

WESTERN CORAL SNAKE

FIVE LINED SKINK

BEHAVIOR Rattlesnakes aren't generally tree climbers and rat snakes are. Box turtles live on land while snapping turtles rarely leave the water. Some lizards jump while others mostly slink. Reptiles have many ways of moving, communicating, and behaving that are specific to a particular animal. Knowing what's normal for the kind of reptile you're identifying helps.

GREEN SNAKE

RANGE & HABITAT This is why field guides have range maps. You can often eliminate possibilities by checking who lives where you are. Sometimes it's the piece that solves the identification puzzle. And don't forget about habitat. That lizard in the desert? It's not a baby alligator.

Talk Like a Reptile Expert

Here are some terms to know when reading field guides, swapping snake stories, and talking turtles.

CARAPACE top shell of turtle

CROSSBAND color band across back and sides of body, but not on belly

DIURNAL active during day

FIELD MARK distinctive feature useful in identification

KEELED SCALES rough, ridged scales

NOCTURNAL active at night

PLASTRON bottom, smaller shell of turtle that covers belly

SCUTES dry, tough plates that make up a turtle's shell

SEMIAQUATIC partially living in water

SHEDS shed skins of reptiles

SOCIAL lives in groups

SOLITARY lives alone except for breeding

URIC PELLET solid or mushy white urine-like waste excreted with feces that reptile bodies make instead of peeing

VENOMOUS able to inject poison by bite (or sting)

VERNAL POND OR POOL small, shallow, temporary water bodies that fill with spring rains and snow-melt, but dry out in summer

What to Do When You Spot a Snake

Most snakes in North America are harmless. Those that are venomous belong to one of the four following groups. Except for coral snakes, all our venomous snakes are pit vipers—snakes with heat-sensing pit organs. Pit vipers have large, wide, flattened, triangular-shaped heads; hinged fangs; eyes with vertical pupils; and pits between the eye and nostril. (Not that you should be close enough to see them!)

PIT VIPER FEATURES

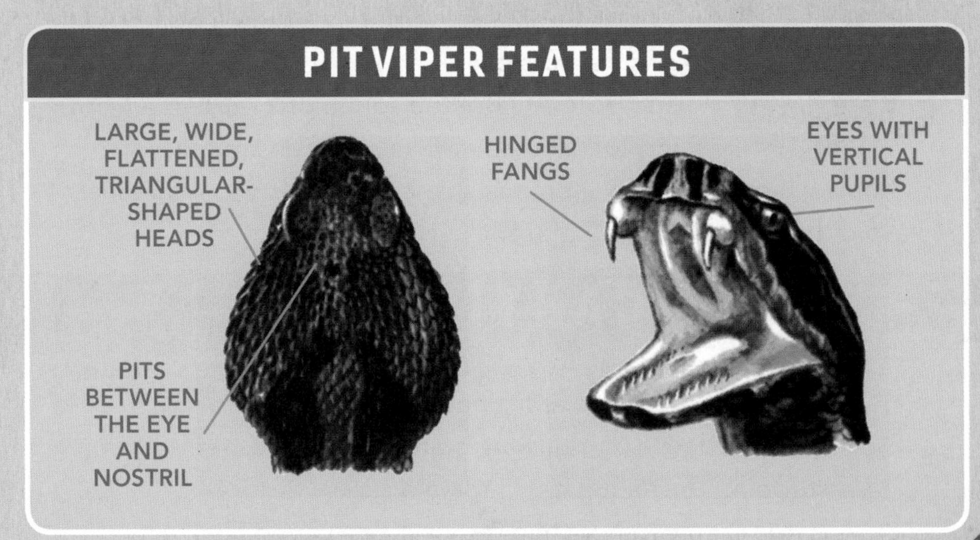

North American Poisonous Snakes

CORAL SNAKES Colorful snakes with black and red sections bordered by yellow stripes. (Not to be confused with harmless king snakes, whose red sections are NOT bordered by yellow stripes. Here's a mnemonic to help you remember which is which: *Red touch yellow (coral snake), kill a fellow. Red touch black (king snake), venom lack.* While coral snake venom is super strong, the fangs that deliver it are small and their owners shy and reclusive.

COPPERHEAD Most common venomous snakebite, but also one of the least potent. Masters of camouflage, they're difficult to see on the forest floor and are quick to strike when they feel threatened.

COTTONMOUTH Also called a water moccasin, it's the only venomous water snake in North America. When threatened, it coils up tight and opens wide, showing off its cottony white mouth lining.

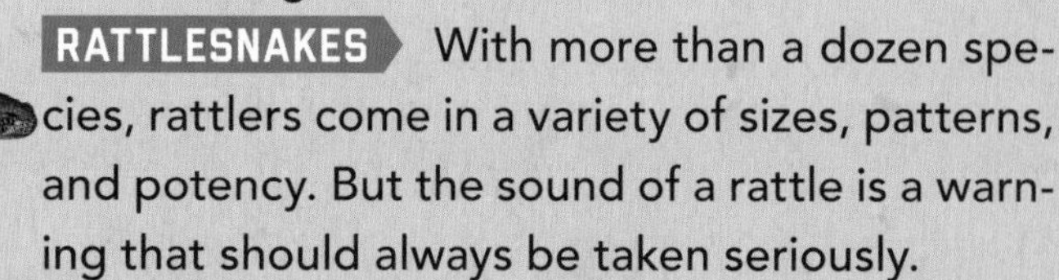

RATTLESNAKES With more than a dozen species, rattlers come in a variety of sizes, patterns, and potency. But the sound of a rattle is a warning that should always be taken seriously.

Know which kinds of venomous snakes live near you—and learn to avoid them. If you're hiking in snake country:

CAUTION!

- Wear boots and loose-fitting long pants.
- Watch where you put hands and feet. Look before stepping over a log or climbing a rock.
- Listen for movement and rattles. Keep your distance.

TRY IT → Spot and Identify a Reptile

Time to try out your reptile recognition skills! Lizards or pond turtles are probably the easiest wild reptiles to find and observe. But if you live in 'gator country or often see snakes in a particular place, that could work, too. Be safe!

WHAT YOU'LL NEED

- binocs

STEP 1 Find a wild reptile. (There are herp hunting tips on page 311.)

STEP 2 Use your binoculars to watch it. As you check it out, pay attention to:

- Shape of its head.
- Scales and claws.
- Color patterns, bands, or lines.
- Whether it moves in a start-stop jerky way or smoothly forward.

STEP 3 What would you say are its *most* distinctive characteristics—its field marks? Rough scales? Blue tail? Really long length? Pointy snout?

STEP 4 Congrats on your spotting!

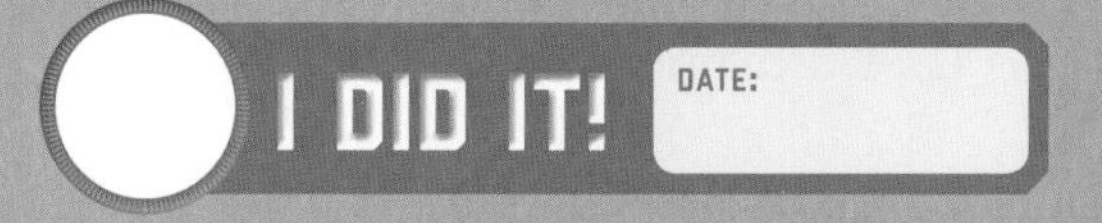

TRACK IT ↘ Tracking Reptiles

Best friends with a reptile now? Record what you observed while watching it.

COLLARED LIZARD

WHAT YOU'LL NEED

- a pencil or pen

STEP 1 Fill out the page and draw the reptile below.

DATE

TIME

LOCATION

WEATHER

Describe habitat

Group ☐ snake ☐ lizard ☐ turtle ☐ alligator

Estimate length of body ....

and tail (if lizard) ....

Do you think it's an adult?

☐ yes ☐ no ☐ not sure

Color

Stripes, spots, or lines?

Its field marks?

What's it doing?

How does it move?

Solitary or in a group?

Describe habitat

Draw the reptile and label its field marks.

STEP 2 Can you identify the reptile? Look for it in the Reptile Identification section (pages 358-381). Make a positive ID? Check off its I SAW IT! box and fill in the blanks. Nice work!

CHAPTER 4

Herp Tracking

Hopefully you've gotten outside, spotted some snakes, and identified a few frogs. But seeing an amphibian or reptile isn't the only way to know there are herps about. All animals leave behind evidence of their actions—not just coyotes and beavers. Learning to recognize some basic herp signs and tracks can clue you in to where to find actual animals. And the signs themselves say a lot about local cold-blooded creatures—if you know how to read them.

Common Herp Tracks

Toads, lizards, and their herp kin make tracks as they hop and scuttle around. And while not all turtle tracks look the same and frog footprints vary by species, you can narrow your search by knowing the characteristics of each herp group's tracks. The overall shape differences between snake, lizard, turtle, frog, toad, and salamander tracks aren't hard to see. Look for these shapes and characteristics:

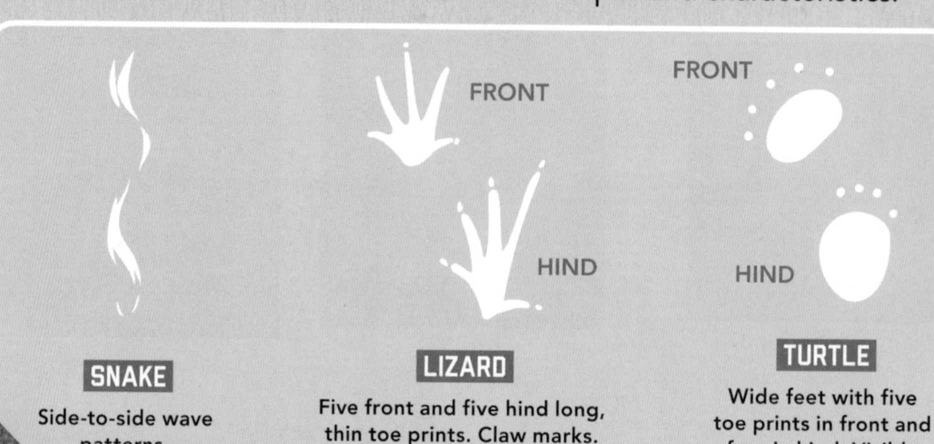

SALAMANDER

Four front and five hind toe prints. Tail drag often present.

FROG

Four front and five hind toe prints. Hind prints webbed. Front print turns inward.

TOAD

Four front and five hind toe prints. Front print turns inward.

Herp Holes, Shells, and Skin Sheds

Tracks aren't the only kind of reptile sign. There are also sheds, the molted skin of snakes, lizards, and turtles. Reptile skin is too tough to stretch as the animal grows bigger. Instead, new expanded skin grows underneath and forces off the old top layer. Skin sheds retain the pattern and spacing of their owners' scales, so they are a great find and can help with identification.

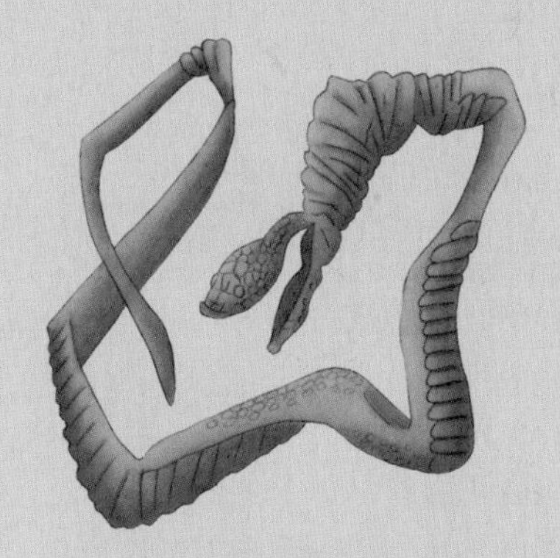

A rat snake shrugged out of this shed.

Turtles shed the outer layer of their shells in scute-shaped flakes.

Burrows dug by tortoises, muddy soaking-holes excavated by box turtles, and even depressions swished out by sleepy snakes are all evidence that a reptile was there, too. Old dug-out lizard, snake, or turtle nests with leathery broken eggshells tell you the eggs hatched. Reptile nests are usually buried, so they're hard to find until everyone has left.

Amphibian Egg Spotting Tips

Amphibian eggs are often easier to spot than reptile eggs—if you know where to look. Search in springtime, and you'll likely come across jellylike egg masses or gooey egg chains in ponds, ditches, or along creek edges. Many species of frogs, toads, and salamanders seek out springtime, or vernal, ponds to deposit their eggs. Why? These temporary ponds lack fish because they dry out in summer—fish are major predators of amphibian eggs. Some kinds of salamanders and frogs travel far and wide to find fish-free places to lay eggs.

How can you tell which amphibian laid the eggs? The eggs of different species vary, but here are some general guidelines for distinguishing frog, toad, and salamander egg masses laid in water. (Some terrestrial salamanders lay eggs under damp rocks and moist logs. Their larvae go through tadpole stage while still inside the egg, then hatch out as fully formed mini salamanders.)

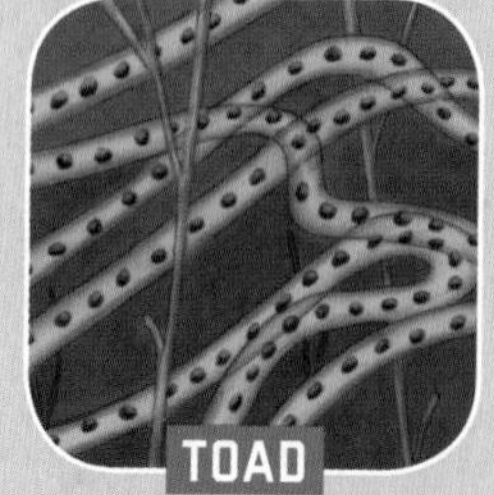

Slimy chain-like strands of dark eggs. Sometimes coiled into clumps.

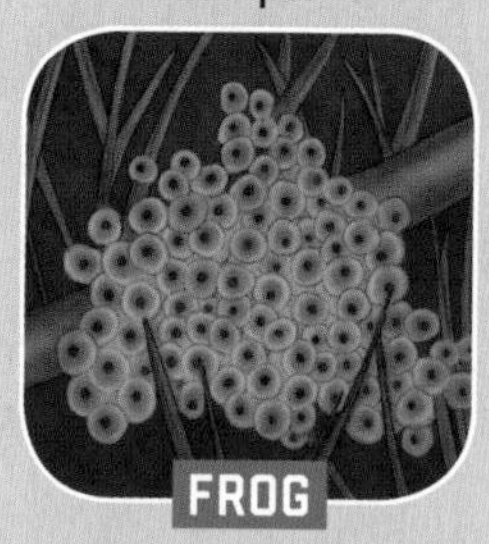

Dark eggs each wrapped in a clear sphere of gel that clump together. Surface of mass shows curve of outer eggs.

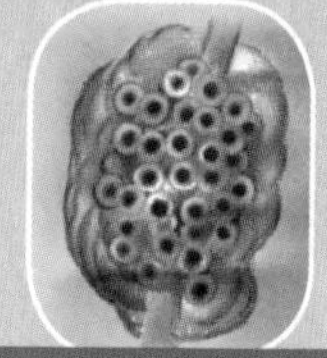

Firm mass with an extra thick layer of clear gel that surrounds all the eggs. Often attached to stick or plant under water's surface.

Amphibian Calls

American toad calling.

Frog and toad calls are nearly as unique to each species as bird songs are. You don't have to see frogs and toads to know they're there—just listen for their calls! During breeding season male frogs and toads move to ponds, ditches, and other watery egg-laying sites. Once in place, the males compete for females by calling to them. Some species are nearly impossible to separate without hearing their calls. The gray tree frog (*Hyla versicolor*) and Cope's gray tree frog (*Hyla chrysoscelis*) look the same. But they sound different. You've got to listen up to know who's who.

Herp Scat Spotting Tips

As we've discussed with birds and mammals, everybody poops, including reptiles and amphibians. Seeing scat is another way to know who's around. Amphibian scat can be hard to find and see. Frogs and salamanders often poop in water or leave scat in damp, dark places. In contrast, reptile scat is often spotted on sunny basking rocks or logs. Unlike mammal scat, reptile poop usually has two parts. A fecal part that's brownish and a white uric pellet. Reptiles only have a single body opening for shuttling out waste, the cloaca. Instead of peeing, urea waste becomes uric acid and is excreted as a white pellet or mushy paste. (That's the same white stuff in bird poop.) It's another way reptiles are well suited to arid environments. They don't pee out water!

WOODRAT
PELLET

TRY IT → Hunt Down Herp Sign

Find a herp sign and try to identify who made it.

WHAT YOU'LL NEED

- a pencil or pen

STEP 1 Take a walk or hike and have a look around! Muddy riverbanks and beaches are great for tracks. Rock piles and walls are good lizard scat locations. If it's springtime, search around ponds and ditches for amphibian eggs.

STEP 2 Check off any herp signs you see or hear:

- ☐ shed snake or lizard skin
- ☐ amphibians' eggs or tadpoles in water
- ☐ herp tracks in sand or mud
- ☐ old nest with eggshells
- ☐ frog or toad calls
- ☐ reptile scat
- ☐ tortoise burrow
- ☐ box turtle holes

STEP 3 Think about who might have made the sign—any ideas?

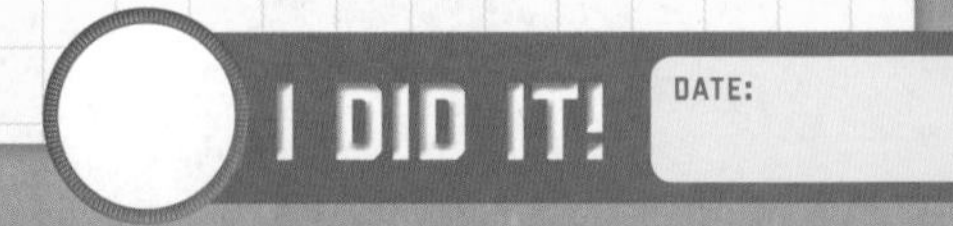

TRACK IT ↘ Tracking Herp Signs

Keeping track of the tracks, scat, eggshells, egg masses, or other herp signs you find is a terrific way to know where to look for herps in the future. Plus, you can investigate if a track-filled route is a regular pathway or just an occasional stop, or whether frogs lay eggs the same time each year.

WHAT YOU'LL NEED

> a ruler, a pencil or pen, a magnifying lens

Found a track? Record it below.

DATE

TIME

LOCATION

WEATHER

Where exactly did you find the track?

How many prints were there?

Draw the track here:

Is this the actual size of the track? ☐ yes ☐ no

If not, write down its length ....

and width ....

Who do you think made the tracks you spotted?

Found a different kind of sign, like scat, an old nest, or eggs? Track it below.

DATE

TIME

LOCATION

WEATHER

What kind of sign did you find?

Where exactly did you find it?

Draw the sign here

Is this its actual size? ☐ **yes** ☐ **no**

If not, write down its length ....

and width ....

Who do you think made the sign you spotted?

I DID IT! DATE:

TRY IT → Who's That Calling?

WESTERN CHORUS FROG

Who's that croaking, trilling, or *ribbiting*? There's nothing more satisfying than recognizing a particular frog's call and then spotting the frog itself. (Aha, I knew it!)

STEP 1 Find some frogs or toads to listen to. Where you're looking and when makes a difference, so ask around. Frogs call most of the year in some parts of Florida, for example. Other places mostly have frogs that call in spring, or after a big rain. Ponds and wetlands with resident frogs might call all summer. Some species of frogs and toads call during the day and others call at night. Dusk to early night is often prime time.

STEP 2 Settle in and listen. It may take a while for the frogs near a pond to forget you're there and start up again after your arrival.

STEP 3 Pick out one particular call that is repeating. Can you tell *where* it's coming from?

STEP 4 Try to find the calling frog or toad. Did you spot it? Use the Amphibian Identification (pages 338-357) and/or a field guide to help identify it. If you got an ID, congratulations! Check off its I SAW IT! box and fill in the blanks.

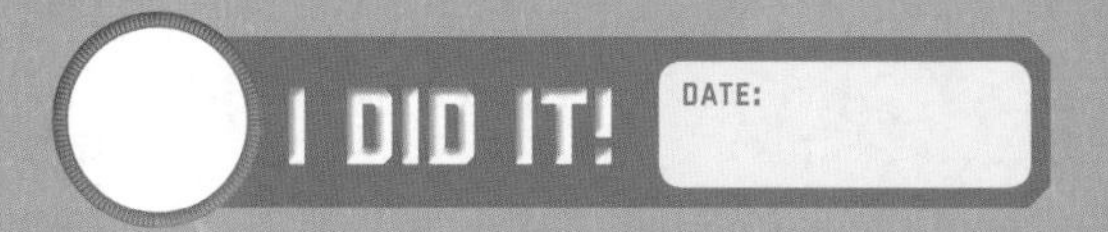

AMPHIBIAN IDENTIFICATION

Welcome to your guide to Amphibian Identification. Here are some tips to get started:

COMMON
FROGS & TOADS

Amphibians are grouped by kind. Common frogs and toads first, then salamanders, which includes newts.

Bumpy, rough
skin.

The labeled main illustration points out field marks. Take note!

Remember, amphibian skin color is variable. Patterns and field marks like stripes, spots, and lines are better clues to identification.

Behavior often tells you when the animal is active, as well as other info including details on frog calls.

Check the range map to see where the amphibian lives.

FROG/TOAD BODY PARTS

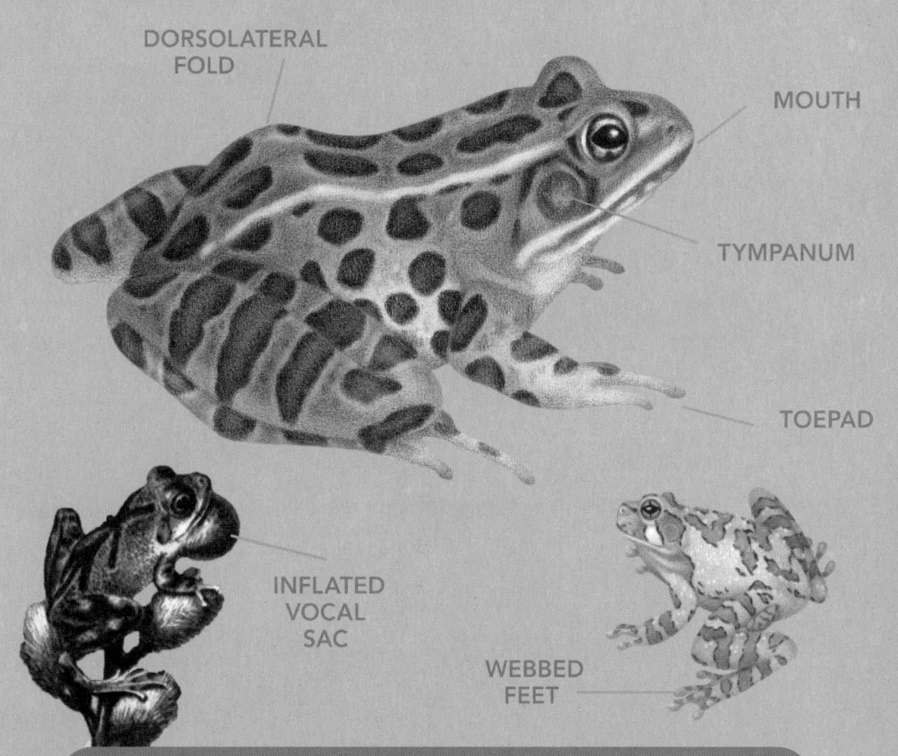

SALAMANDER BODY PARTS

MOIST, SCALELESS SKIN

CLAWLESS TOES

MOUTH

COSTAL GROOVES

Green Frog

(Lithobates clamitans)

SIZE 2–3 inches (5–7½ cm) snout to rump

SHAPE Medium-size frog with webbed hind feet, dual ridges down its back, and large, bulging eyes.

COLOR Green to bronzy-brown body with dark spots on its back and a lighter green upper lip. Males have yellowish throats.

FEMALE

BEHAVIOR Aquatic. Males call a single note that sounds like a low, banjo-like twang.

HABITAT Ponds, swamps, and shallow streams.

POINT OF FACT The ridges along its back help to tell it apart from a larger bullfrog.

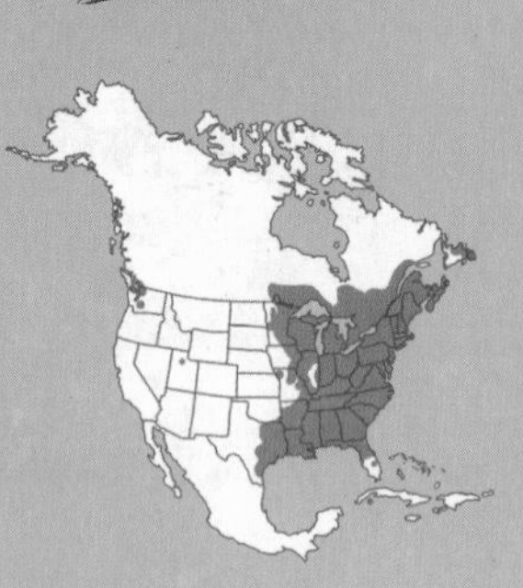

I SAW IT!

WHEN I SAW IT
DATE

WHERE I SAW IT
SPECIFIC LOCATION, INCLUDE STATE

WHAT IT WAS DOING

NOTES

American Toad

(Anaxyrus americanus)

SIZE 2–3½ inches (5–9 cm) snout to rump

SHAPE Plump, medium-size toad with short legs, big eyes, and bumpy skin.

COLOR Mixes of brown, gray, red, and tan with dark spots and colored patches.

BEHAVIOR Mostly nocturnal. It lives mostly on land, except during breeding season when males trill a long, cricket-like call from water-filled ditches, along creeks, and near pools of water.

HABITAT Forests, fields, backyards, parks.

POINT OF FACT It catches insects with its lightning-quick tongue. Zap!

I SAW IT!

WHEN I SAW IT

DATE

WHERE I SAW IT

SPECIFIC LOCATION, INCLUDE STATE

WHAT IT WAS DOING

NOTES

Woodhouse's Toad

(Anaxyrus woodhousii)

SIZE 2½–4 inches (6½–10 cm) snout to rump

SHAPE Medium-size, plump toad with a pointed snout and bumpy skin.

COLOR Green-brown to gray variations with dark blotches on its back.

BEHAVIOR Nocturnal. It lives on land, except when it's breeding.

HABITAT River valleys, mountain canyons, grasslands, desert, farm fields.

POINT OF FACT Males make a one to three second *waaaaah* call near water to attract mates who lay eggs in ponds, lakes, watery ditches, and even livestock watering tanks.

I SAW IT!

WHEN I SAW IT

DATE

WHERE I SAW IT

SPECIFIC LOCATION, INCLUDE STATE

WHAT IT WAS DOING

NOTES

Great Plains Toad

(Anaxyrus cognatus)

SIZE 2–3½ inches (5–9 cm) snout to rump

SHAPE Medium-size, beefy toad with bumpy skin and large eyes.

COLOR Gray to green with a pattern of large, dark, blotchy spots outlined with a light green-gray.

BEHAVIOR Nocturnal. It spends the daytime underground and hunts insects at night.

HABITAT Grasslands, farm fields, scrublands.

POINT OF FACT It eats many garden and crop pests, so it's valuable to farmers.

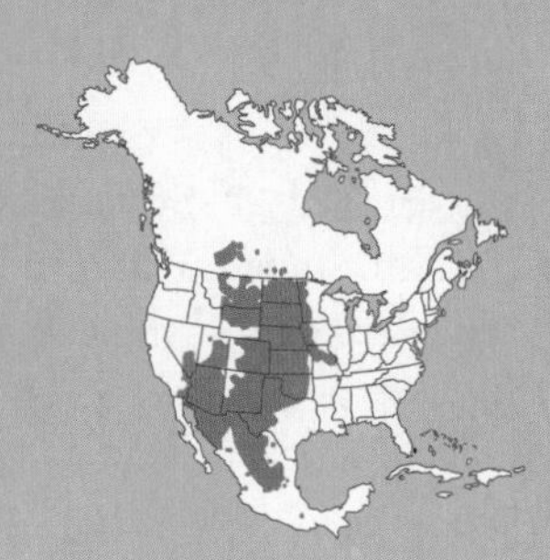

I SAW IT!

WHEN I SAW IT

DATE

WHERE I SAW IT

SPECIFIC LOCATION, INCLUDE STATE

WHAT IT WAS DOING

NOTES

Eastern Spadefoot

(Scaphiopus holbrookii)

SIZE 1¾–2¼ inches (4½–5½ cm) snout to rump

SHAPE Smallish toad with smooth skin, huge, bulging eyes, and a hard, hoof-like "spade" on the inside of each hind foot for digging burrows.

COLOR Dark green-brown to black with yellow lines down its back behind each eye.

BEHAVIOR It spends much of its life hibernating in an underground burrow. Heavy rainfall draws it out of its burrow to breed.

HABITAT Forests and fields with soft, loose, and sandy soil.

POINT OF FACT Some people are allergic to spadefoots and get a runny nose, watery eyes, and fits of sneezing when handling them.

I SAW IT!

WHEN I SAW IT

DATE

WHERE I SAW IT

SPECIFIC LOCATION, INCLUDE STATE

WHAT IT WAS DOING

NOTES

Gray Tree Frog

(Hyla versicolor)

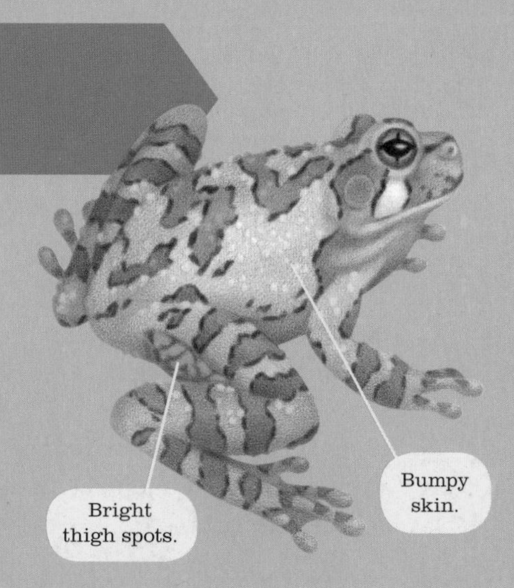

SIZE 1¼–2¼ inches (3–5½ cm) snout to rump

SHAPE Small, squat tree frog with bumpy skin and large toepads.

COLOR Gray to brown with a mottled, lichen-like pattern on its back and yellow-orange spots on the insides of its thighs.

BEHAVIOR Nocturnal. It lives and hunts for insects in low trees and bushes, and only comes to the ground to breed.

POINT OF FACT When the air temperature of late spring evenings rises above 59°F (15°C), males sweetly trill to mates from bushes near water.

I SAW IT!

WHEN I SAW IT

DATE

WHERE I SAW IT

SPECIFIC LOCATION, INCLUDE STATE

WHAT IT WAS DOING

NOTES

....

....

Pacific Tree Frog

(Pseudacris regilla)

SIZE ¾–2 inches (2–5 cm) snout to rump

SHAPE Small, slender tree frog with long toes and round toepads.

COLOR Can be green to tan or red to brown with dark lines through each eye. Males have gray throats.

BEHAVIOR Hangs out in shrubs near the ground and in grasses near the water.

HABITAT Forests, grassland, farmland, shrubby areas near water.

POINT OF FACT Many California moviemakers use this common frog's *ribbit, ribbit* call in soundtracks—regardless of where the scene is filmed.

I SAW IT!

WHEN I SAW IT
DATE

WHERE I SAW IT
SPECIFIC LOCATION, INCLUDE STATE

WHAT IT WAS DOING

NOTES

Spring Peeper

(Pseudacris crucifer)

SIZE ¾–1¼ inches (2–3 cm) snout to rump

SHAPE Small frog with a wide head and large toepads.

COLOR Tan to brown with a dark, X-shaped mark on its back.

BEHAVIOR Nocturnal. It hunts near the ground for spiders, ants, and other small insects.

HABITAT Ponds and swamps in forests, moist woods.

POINT OF FACT It's named for its high-pitched, repeating *peep, peep* trill, which is often the first frog call of spring.

I SAW IT!

WHEN I SAW IT

DATE

WHERE I SAW IT

SPECIFIC LOCATION, INCLUDE STATE

WHAT IT WAS DOING

NOTES

Western Chorus Frog

(Pseudacris triseriata)

SIZE ¾–1¼ inches (2–3 cm) snout to rump

SHAPE Small, slender frog with smooth skin, and slightly webbed toes with tiny pads.

COLOR Green-gray to brown with three dark stripes down its back and stripes through its eyes.

BEHAVIOR Nocturnal. It hunts mosquitoes, ants, small beetles, flies, and spiders near temporary ponds, wet meadows, grassy pools, and river swamps.

HABITAT Grassy swamps, wet woods, farmland, neighborhoods.

POINT OF FACT Females lay their sticky eggs in clusters onto underwater plants for protection.

I SAW IT!

WHEN I SAW IT

DATE

WHERE I SAW IT

SPECIFIC LOCATION, INCLUDE STATE

WHAT IT WAS DOING

NOTES

Wood Frog

(Lithobates sylvaticus)

SIZE 1¼–2¾ inches (3–7 cm) snout to rump

SHAPE Medium-size frog with back ridges, large eyes, and long back legs.

COLOR Tan to copper mottled brown with a black mask across its eyes and cheeks.

BEHAVIOR It emerges from hibernation in February or March to gather at cold ponds and pools where it breeds and lays eggs over a few days. Then it returns to the woods. Its call is a hoarse, duck-like *quack, quack*.

HABITAT Woodlands, grasslands, and tundra.

POINT OF FACT It's sometimes called a "robber's mask frog" because of the stripe pattern on its face.

I SAW IT!

WHEN I SAW IT
DATE

WHERE I SAW IT
SPECIFIC LOCATION, INCLUDE STATE

WHAT IT WAS DOING

NOTES

....

....

Northern Leopard Frog

(Lithobates pipiens)

SIZE 2–3½ inches (5–9 cm) snout to rump

SHAPE Medium-size, lean frog with a rounded snout, a large mouth, and deep back ridges.

COLOR Brown to green variations with rows of large spots on its back and legs.

BEHAVIOR It hunts a wide variety of prey by day, including spiders, insects, slugs, earthworms, and smaller frogs.

HABITAT Slow streams, marshes, bogs, ponds, canals, lakes, and springs.

POINT OF FACT It's the frog most commonly dissected in biology classes.

I SAW IT!

WHEN I SAW IT

DATE

WHERE I SAW IT

SPECIFIC LOCATION, INCLUDE STATE

WHAT IT WAS DOING

NOTES

Bullfrog

(Lithobates catesbeianus)

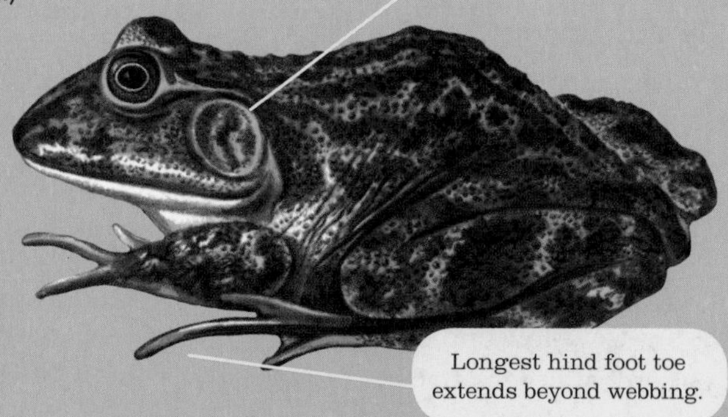

SIZE 3½–6 inches (9–15 cm) snout to rump

SHAPE Large frog with big, bulging eyes, a huge mouth, and thick legs.

COLOR Green, but can sometimes have brown mottling with gold eyes in the Southeast.

BEHAVIOR Aquatic. It's active during warm, wet weather, no matter the time of day. It sits and waits for prey to pass, grabbing it with its speedy tongue. Gulp!

HABITAT Lakes, ponds, swamps, marshes, bogs, and slow-moving streams.

POINT OF FACT It becomes an invasive species where it's introduced because it gobbles up and displaces native frogs.

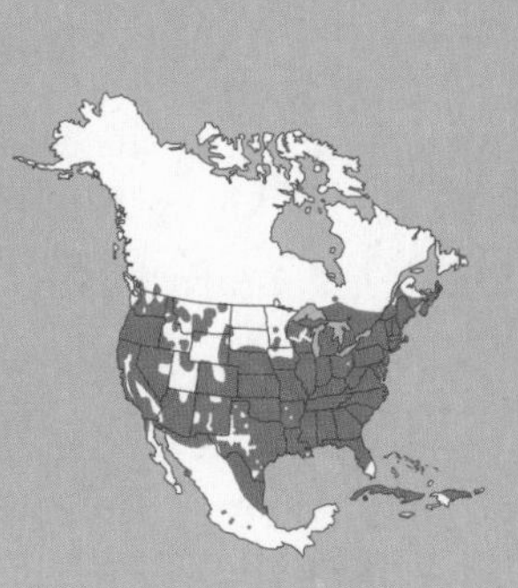

I SAW IT!

WHEN I SAW IT
DATE

WHERE I SAW IT
SPECIFIC LOCATION, INCLUDE STATE

WHAT IT WAS DOING

NOTES

COMMON SALAMANDERS

Rough-skinned Newt

(Taricha granulosa)

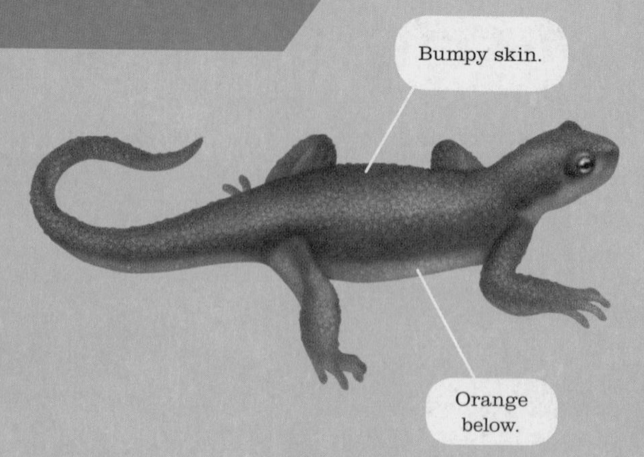

SIZE 5–8½ inches (12½–21½ cm) snout to tail tip

SHAPE Large and thick newt with small eyes and bumpy skin.

COLOR Brown and black on top with yellow and orange underneath on its belly.

BEHAVIOR Diurnal. When threatened, it arches its head toward its curled-up tail to show off its colorful undersides in warning.

HABITAT Ponds, streams, moist woods, and grasslands.

POINT OF FACT The strong toxin in its skin is powerful enough to kill most predators, but garter snakes are completely immune to it.

I SAW IT!

WHEN I SAW IT
DATE

WHERE I SAW IT
SPECIFIC LOCATION, INCLUDE STATE

WHAT IT WAS DOING

NOTES

Eastern Newt

(Notophthalmus viridescens)

Spots.

ADULT

Light belly.

JUVENILE

SIZE 2¼–5 inches (5½–12½ cm) snout to tail tip

SHAPE Medium-size newt with extreme differences in its adult and juvenile forms.

COLOR Adults have olive-green backs, yellow bellies, red spots with black outlines on its back, and small black spots all over. Juveniles (efts) are a red-orange color with black-outlined spots on their backs.

Juvenile eft lives on land for up to five years.

BEHAVIOR Aquatic larvae born in springtime change into land-living efts in the fall. Efts live in the woods, hibernating under logs and rocks in winter, for two to five years until they mature into aquatic adults and move back to water. After their births in water, young newts become efts and leave for land at age two. Once mature, they return to the water to live and breed.

HABITAT Adults live in ponds, small lakes, marshes, and quiet streams. Efts live in moist woods.

POINT OF FACT It can live as long as fifteen years, and the bad-tasting toxin its skin oozes keeps predators away.

I SAW IT!

WHEN I SAW IT

DATE

WHERE I SAW IT

SPECIFIC LOCATION, INCLUDE STATE

WHAT IT WAS DOING

NOTES

Tiger Salamander

(Ambystoma tigrinum)

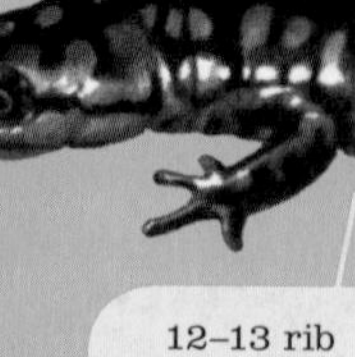

SIZE 7–8¼ inches (18–21 cm) snout to tail tip

SHAPE Large, thick salamander with a big head and a wide, rounded snout.

COLOR Black with yellow spots or blotches all over. Belly greenish yellow with dark mottling.

BEHAVIOR Spends much of the year in an underground burrow, leaving only in wet weather to breed and find food.

HABITAT Deciduous and mixed forests, grasslands.

POINT OF FACT It's the largest land-living salamander in North America—with a record size of 13 inches (33 cm) long! It's also the most wide-ranging salamander.

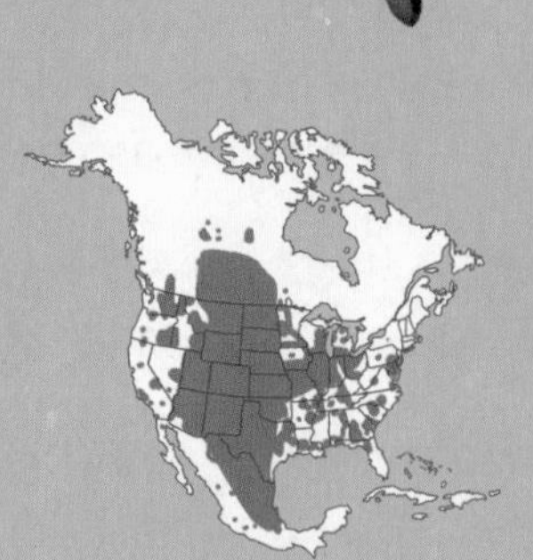

I SAW IT!

WHEN I SAW IT

DATE

WHERE I SAW IT

SPECIFIC LOCATION, INCLUDE STATE

WHAT IT WAS DOING

NOTES

COMMON SALAMANDERS

Spotted Salamander

(Ambystoma maculatum)

SIZE 4½–8 inches (11½–20½ cm) snout to tail tip

SHAPE Large, stout salamander with a wide, rounded snout and grooved ribs.

COLOR Dark gray to black with round, yellow spots in rows down its back.

BEHAVIOR Nocturnal. It stays underground except to forage for food on rainy nights or during breeding season.

HABITAT Deciduous and mixed forests near swamps, streams, and spring ponds.

POINT OF FACT Glands on its back and tail release a sticky white toxin when it's threatened.

I SAW IT!

WHEN I SAW IT

DATE

WHERE I SAW IT

SPECIFIC LOCATION, INCLUDE STATE

WHAT IT WAS DOING

NOTES

Red-backed Salamander

(Plethodon cinereus)

SIZE 2¼–4 inches (5½–10 cm) snout to tail tip

SHAPE Long and skinny salamander with bulging eyes and shiny skin.

COLOR Dark, glossy gray to black variations, sometimes with a yellow to red stripe down its back. It also has a light belly with dark mottling.

BEHAVIOR Terrestrial hunter of mites and other small bugs that live in leaf litter and under rocks and logs.

HABITAT Forests with moist soil and leaf litter layers.

POINT OF FACT It's a lungless salamander that breathes through its skin and the mucus-covered lining of its mouth and throat.

I SAW IT!

WHEN I SAW IT
DATE

WHERE I SAW IT
SPECIFIC LOCATION, INCLUDE STATE

WHAT IT WAS DOING

NOTES

Common Mudpuppy

(Necturus maculosus)

SIZE 8–13 inches (20½–33 cm) snout to tail tip

SHAPE Very large and long salamander with tall tail fin, feathery gills, tiny eyes, and slime-covered skin.

COLOR Gray to brown with dark spots and red gills.

BEHAVIOR Aquatic and nocturnal. It moves along streams and lake bottoms to hunt crayfish, worms, snails, and fish.

HABITAT Lakes, ponds, rivers, streams.

POINT OF FACT It makes a squeaky noise that doesn't sound very puppy-like considering its name, but it does have teeth and it can bite!

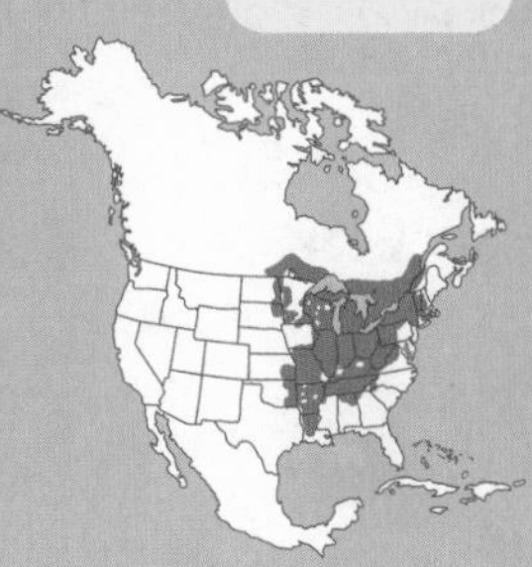

I SAW IT!

WHEN I SAW IT

DATE

WHERE I SAW IT

SPECIFIC LOCATION, INCLUDE STATE

WHAT IT WAS DOING

NOTES

....

....

REPTILE IDENTIFICATION

Welcome to your guide to Reptile Identification. Here are some tips to get started:

COMMON SNAKES

Reptiles are grouped by kind. Common snakes first, then lizards, next turtles, and the alligator.

Black necklace around head and neck.

The labeled main illustration points out field marks. Take note!

Reptile scale color can vary with age and region—especially snakes. Patterns and field marks like crossbands, stripes, and spots are key clues to identification.

Behavior often tells you when the animal is active, as well as what you might see the reptile doing.

Check the range map to see where the reptile lives.

TURTLE BODY PARTS

LIZARD BODY PARTS

SNAKE PATTERNS

blotches

crossbands

diamonds

rings

stripes

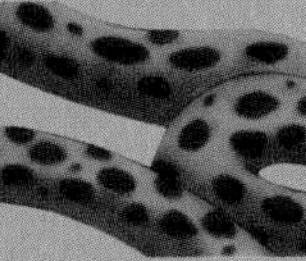
spots

Eastern Hog-nosed Snake

(Heterodon platirhinos)

Upturned snout.

Thick neck.

SIZE 20–33 inches (51-84 cm) in length

SHAPE Thick snake with wide head.

COLOR Its color varies, but usually it's brown to gray with dark splotches or totally black. It's lighter under its tail, and it has a mottled belly.

BEHAVIOR When threatened, it acts like a cobra by puffing up, hissing, and flattening its neck.

Some are patterned, but others are nearly solid black.

HABITAT Open woodlands and edges, hillsides, fields, and places with sandy soil.

POINT OF FACT It avoids being eaten by playing dead—rolling onto its back, shaking, and then lying still with its mouth open and tongue out.

I SAW IT!

WHEN I SAW IT
DATE

WHERE I SAW IT
SPECIFIC LOCATION, INCLUDE STATE

WHAT IT WAS DOING

NOTES

COMMON SNAKES

Prairie Rattlesnake

(Crotalus viridis)

SIZE 35–45 inches (89–114½ cm) in length

SHAPE Heavy-bodied snake with a wide, triangular head and a rattle at the end of its tail.

COLOR Its color varies, but usually it has a brown to green-gray body with dark brown, oval-shaped splotches. Its body darkens to black near its rattle.

BEHAVIOR It kills small mammals and birds with a venomous bite delivered by its fangs.

HABITAT Grasslands, prairies, deserts, pine forests.

POINT OF FACT It spends the winter hibernating in dens, sometimes with large groups of other rattlesnakes.

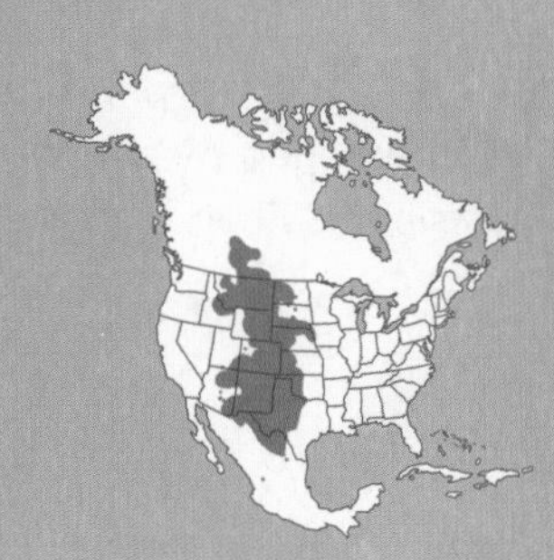

I SAW IT!

WHEN I SAW IT

DATE

WHERE I SAW IT

SPECIFIC LOCATION, INCLUDE STATE

WHAT IT WAS DOING

NOTES

Gopher Snake

(Pituophis catenifer)

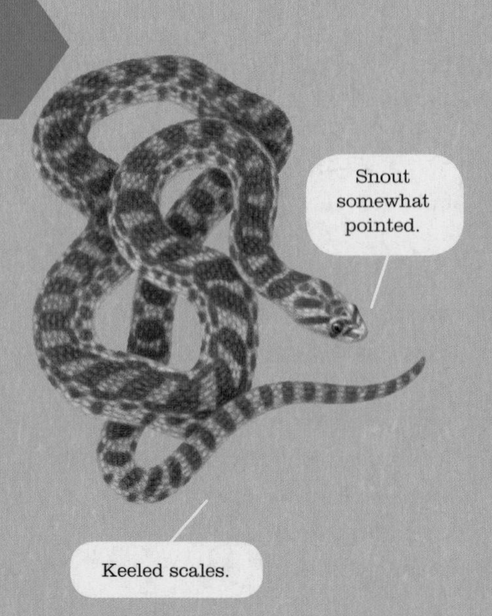

SIZE 50–100 inches (127–254 cm) in length

SHAPE Large, beefy snake with a small head and large eyes.

COLOR Its color varies, but usually it has brown to black blotches on a tan to gray body.

BEHAVIOR Diurnal hunter that kills with constriction by coiling its body around its prey until it suffocates.

HABITAT Woodlands, deserts, farmlands, grasslands, shrublands, forest edges.

POINT OF FACT It pretends to be a rattlesnake when it's threatened by shaking its tail, hissing loudly, and by using its head to strike.

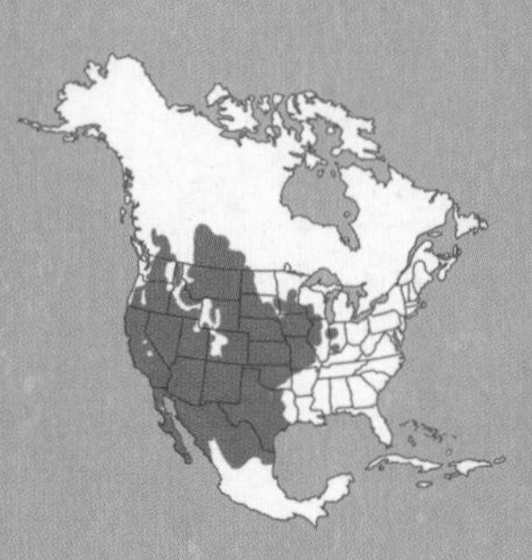

I SAW IT!

WHEN I SAW IT
DATE

WHERE I SAW IT
SPECIFIC LOCATION, INCLUDE STATE

WHAT IT WAS DOING

NOTES

Ringneck Snake

(Diadophis punctatus)

SIZE 10–18 inches (25½–45½ cm) in length

SHAPE Small, thin, and smooth snake.

COLOR Black to green upper body with a yellow to red belly and neck ring.

BEHAVIOR Nocturnal, timid and meek. It hides under stones and leaf litter.

HABITAT Deciduous and mixed forests, grasslands, desert, stream sides.

POINT OF FACT When threatened, it curls up its tail and exposes its bright belly to its enemies.

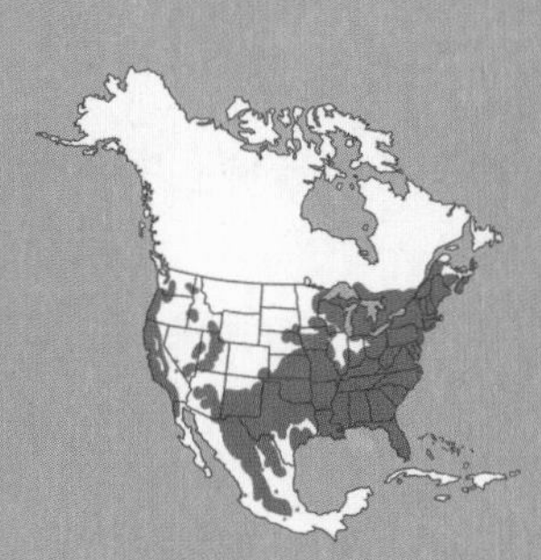

I SAW IT!

WHEN I SAW IT

DATE

WHERE I SAW IT

SPECIFIC LOCATION, INCLUDE STATE

WHAT IT WAS DOING

NOTES

....

....

Common King Snake

(Lampropeltis getula)

SIZE 36–48 inches (91½–122 cm) in length

SHAPE Large snake with smooth scales.

COLOR Its color varies, but usually it has a shiny, dark brown to black body with cream bands or chain-link patterns.

BEHAVIOR A daytime constrictor that hunts rodents, snakes, and other prey.

HABITAT Dry woodlands, rocky hillsides, deserts, swamp and marsh edges.

POINT OF FACT It hunts other snakes, including rattlers! It's immune to pit viper venom. It's truly the king of snakes, indeed!

I SAW IT!

WHEN I SAW IT

DATE

WHERE I SAW IT

SPECIFIC LOCATION, INCLUDE STATE

WHAT IT WAS DOING

NOTES

Coachwhip

(Masticophis flagellum)

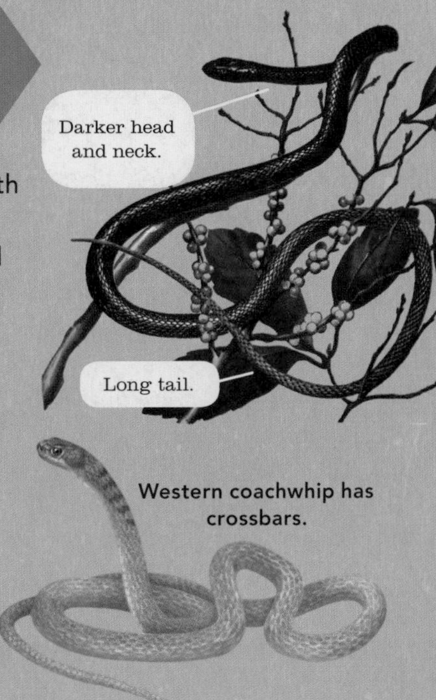

SIZE 42–60 inches (106½-152½ cm) in length

SHAPE Large, slender snake with a long tail and large eyes.

COLOR Its color varies. In the East, it has a dark brown head that fades to a lighter body. In the West, it's tan with dark crossbars on its neck.

Western coachwhip has crossbars.

BEHAVIOR When scared, it disappears in an instant into a burrow or under a rock. It raises its head above grass or rock piles to look around for predators—or prey!

HABITAT Grasslands, dry pastures, open woods.

POINT OF FACT It's one of the fastest snakes in North America, and its name comes from its tail's resemblance to a braided whip.

I SAW IT!

WHEN I SAW IT

DATE

WHERE I SAW IT

SPECIFIC LOCATION, INCLUDE STATE

WHAT IT WAS DOING

NOTES

....

....

COMMON SNAKES

Rat Snake

(Pantherophis)

SIZE 42–72 inches (106½–183 cm) in length

SHAPE Large, thick snake with a square shape like a bread loaf.

COLOR Its color varies. In the North, it's black. In the South, it's yellow or gray with blotches. Coastal Southeast snakes are yellow with stripes.

BEHAVIOR It climbs trees and hangs from branches and rafters. It squirts out a foul-smelling musk when it's threatened.

HABITAT Farm fields, swamps, deciduous forests, old barns and buildings.

POINT OF FACT It can shimmy up a wide tree trunk by grabbing onto the bark.

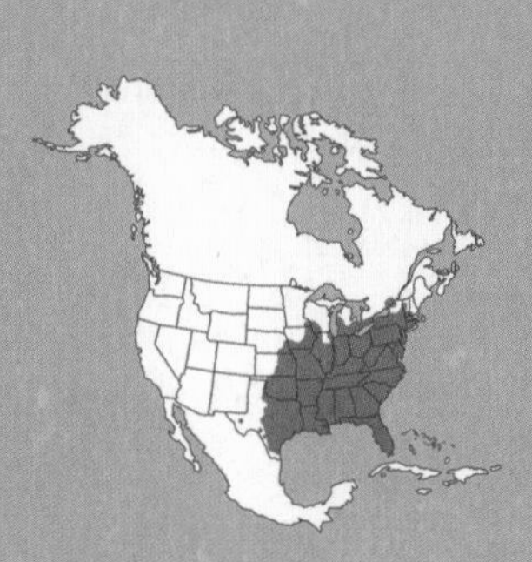

I SAW IT!

WHEN I SAW IT

DATE

WHERE I SAW IT

SPECIFIC LOCATION, INCLUDE STATE

WHAT IT WAS DOING

NOTES

Northern Water Snake

(Nerodia sipedon)

SIZE 22–41 inches (56–104 cm) in length

SHAPE Thick snake with rough skin and an oval head.

COLOR Usually a red-brown to black body with dark neck bands and blotches.

BEHAVIOR Hunts prey by swimming in streams and swamps. It basks on logs in water, on rocky outcrops, and sandy shores.

HABITAT Ponds, lakes, streams, rivers, marshes, swamps.

POINT OF FACT It has long teeth that are good for holding fish and slippery frogs.

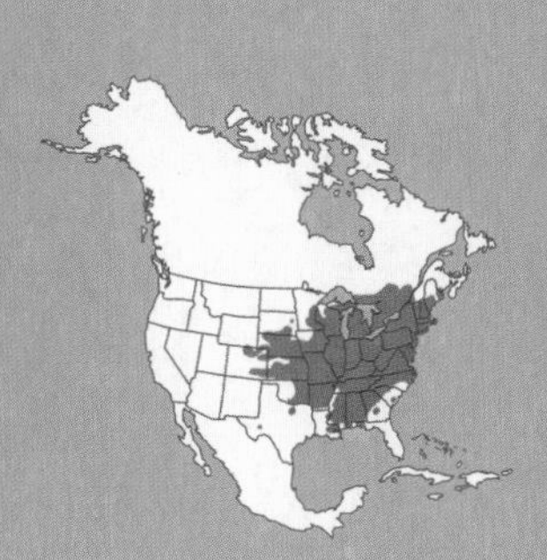

I SAW IT!

WHEN I SAW IT
DATE

WHERE I SAW IT
SPECIFIC LOCATION, INCLUDE STATE

WHAT IT WAS DOING

NOTES

Common Garter Snake

(Thamnophis sirtalis)

Three stripes down back.

SIZE 18–26 inches (45½–66 cm) in length

SHAPE Medium, lean snake.

COLOR Its color varies, but usually it has long stripes down its dark body. It also is sometimes speckled or checkered on its sides.

BEHAVIOR Diurnal hunter of worms, amphibians, insects, and mice. When threatened, it flattens out its body, squeezes out a stinky musk, and strikes the enemy.

CHECKERED MORPH

HABITAT Parks, neighborhoods, farms, gardens, woodlands, near water.

POINT OF FACT It's one of the most commonly seen snakes in the United States and Canada.

I SAW IT!

WHEN I SAW IT

DATE

WHERE I SAW IT

SPECIFIC LOCATION, INCLUDE STATE

WHAT IT WAS DOING

NOTES

COMMON LIZARDS

Collared Lizard

(Crotaphytus collaris)

SIZE 8–14 inches (20½–35½ cm) snout to tail tip

SHAPE Medium-size lizard with an oversize head, a long tail, long hind legs, and small scales.

COLOR Males are yellow-green to brown with lighter spots and dark crossbands. Females are dull but develop red spots on their sides when carrying eggs.

BEHAVIOR It sprints by running on its long back legs with its tail lifted. It looks like a mini T-rex!

HABITAT Dry, rocky hillsides, canyons, scrublands.

POINT OF FACT During their breeding season, males become brightly colored with aqua and yellow crossbands.

I SAW IT!

WHEN I SAW IT

DATE

WHERE I SAW IT

SPECIFIC LOCATION, INCLUDE STATE

WHAT IT WAS DOING

NOTES

Eastern Fence Lizard

(Sceloporus undulatus)

SIZE 4–7½ inches (10–19 cm) snout to tail tip

SHAPE Small lizard with spiny scales, short legs, and small eyes.

COLOR Gray to brown with a blue belly and throat patches, which are faint on females.

BEHAVIOR It's a fast tree climber, and it sunbathes on wood fences, tree stumps, and rocks.

HABITAT Open pine woods, brushlands, prairies, grassy dunes, old barns, and empty buildings.

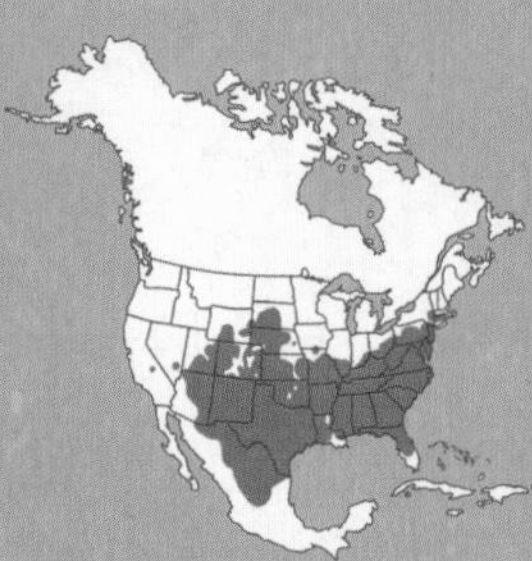

POINT OF FACT It escapes up tree trunks, circling higher and higher while stopping on opposite sides to see if the enemy has left.

I SAW IT!

WHEN I SAW IT
DATE

WHERE I SAW IT
SPECIFIC LOCATION, INCLUDE STATE

WHAT IT WAS DOING

NOTES

Five-lined Skink

(Plestiodon fasciatus)

ADULT

Five stripes.

Narrow snout.

Thick neck.

JUVENILE

SIZE 5–8½ inches (12½–21½ cm) snout to tail tip

SHAPE Small, sleek lizard with a long tail, a thick neck, smooth scales, and a narrow snout.

COLOR Shiny black with five yellow stripes down its body. Juvenile has blue tail.

BEHAVIOR Diurnal. It hunts insects with quick movements among fallen logs and in gardens.

HABITAT Moist woods, gardens, deciduous forests, parks.

POINT OF FACT Skinks lose their tails easily because they sacrifice them to escape threats. However, their tails grow back!

I SAW IT!

WHEN I SAW IT
DATE

WHERE I SAW IT
SPECIFIC LOCATION, INCLUDE STATE

WHAT IT WAS DOING

NOTES

COMMON LIZARDS

Green Anole

(Anolis carolinensis)

Throat fan.

Large toepads.

SIZE 5–8 inches (12½–20½ cm) snout to tail tip

SHAPE Slender, mid-size lizard with a thin tail, a long, pointed snout, and a throat fan.

COLOR Green to brown body with a white to pink throat fan.

BEHAVIOR Climbs trees, bushes, fences, and buildings. It quickly changes its color between green to brown to regulate its temperature and communicate its mood.

HABITAT Parks, neighborhoods, palm fronds, thickets of vines, and bushes.

POINT OF FACT Males chase each other and display to females by bobbing their heads and by doing pushups.

I SAW IT!

WHEN I SAW IT

DATE

WHERE I SAW IT

SPECIFIC LOCATION, INCLUDE STATE

WHAT IT WAS DOING

NOTES

Greater Short-horned Lizard

(Phrynosoma hernandesi)

SIZE 2½–5 inches (6½–12½ cm) snout to tail tip

SHAPE Wide, flat, round body, a large head, and a stubby tail covered in armor-like scales.

COLOR Gray to red brown with a mottled pattern.

BEHAVIOR Spends the day hunting ants and other small insects. It uses a side-shuffling motion to bury itself in soil at night.

HABITAT Shortgrass prairie, mountain meadows, arid woodlands, desert.

POINT OF FACT It squirts blood from its eyes when it's threatened.

I SAW IT!

WHEN I SAW IT
DATE

WHERE I SAW IT
SPECIFIC LOCATION, INCLUDE STATE

WHAT IT WAS DOING

NOTES

....

....

Eastern Box Turtle

(Terrapene carolina)

Domed, colorful carapace.

SIZE Shell length 4½–6 inches

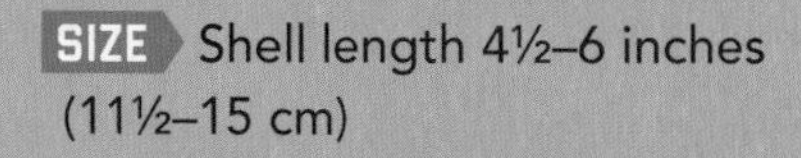

(11½–15 cm)

SHAPE Medium-size turtle with a high-domed carapace, a hinged plastron, scaly legs, and a beaked mouth.

COLOR Brown-black body with yellow and orange markings on its carapace and skin. Male eyes are usually red, and female eyes are yellow.

BEHAVIOR Land-living terrestrial turtle that soaks in cool mud or puddles on hot days.

HABITAT Woodlands, wet meadows, old fields, thickets.

POINT OF FACT The hinge on its plastron allows for a total withdrawal into its shell, protecting it from hungry predators.

I SAW IT!

WHEN I SAW IT

DATE

WHERE I SAW IT

SPECIFIC LOCATION, INCLUDE STATE

WHAT IT WAS DOING

NOTES

Red-eared Slider

(Trachemys scripta elegans)

Red or yellow thick stripe behind eye.

SIZE Shell length 5–8 inches (12½–20½ cm)

SHAPE Small water turtle with an oval-shaped, rough carapace with sharp edges and clawed, webbed feet.

COLOR Dark green carapace with yellow streaks and bars, and a yellow plastron with green markings. Older adults are often darker and have fewer stripes.

BEHAVIOR Aquatic. It hauls out of shallow, quiet waters onto logs to bask in the sun.

HABITAT Slow rivers and streams, ponds, lakes, swamps.

POINT OF FACT Millions of red-eared sliders were exported around the world as pets. Non-native populations now live in Australia, Southeast Asia, Europe, and Israel. Remember, never set pets free in the wild!

I SAW IT!

WHEN I SAW IT

DATE

WHERE I SAW IT

SPECIFIC LOCATION, INCLUDE STATE

WHAT IT WAS DOING

NOTES

Common Snapping Turtle

(Chelydra serpentina)

SIZE Shell length 8–14 inches (20½–35½ cm)

SHAPE Large, beefy, fleshy, and thick turtle with a big head, a long tail, and thick claws.

COLOR Muddy brown to gray.

BEHAVIOR Aquatic. It hides in river and pond mud to ambush fish, frogs, snakes, birds, and even muskrats.

HABITAT Shallow ponds, rivers, coastal marshes.

POINT OF FACT It's unable to retreat into its shell, so it uses its bone-snapping jaws as its defense. Watch out!

I SAW IT!

WHEN I SAW IT

DATE

WHERE I SAW IT

SPECIFIC LOCATION, INCLUDE STATE

WHAT IT WAS DOING

NOTES

Painted Turtle

(Chrysemys picta)

SIZE Shell length 4½–9 inches (11½–23 cm)

SHAPE Medium-size turtle with a smooth, oval carapace, small legs and tail, and an unhinged plastron.

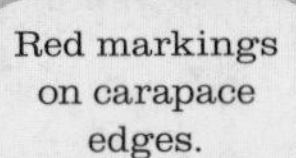

COLOR Dark gray-green to black carapace with red markings along the edge, a yellow to dark green head, and a yellow plastron with splotchy markings.

BEHAVIOR Aquatic. It basks on logs, and sometimes lies on top of another turtle.

HABITAT Marshes, ponds, slow creeks, lake edges.

POINT OF FACT Females are larger and they dig nests in soil near water to lay and cover up clutches of oval-shaped eggs.

I SAW IT!

WHEN I SAW IT

DATE

WHERE I SAW IT

SPECIFIC LOCATION, INCLUDE STATE

WHAT IT WAS DOING

NOTES

Spiny Softshell

(Apalone spinifera)

SIZE Shell length 5–21 inches (12½–53½ cm)

SHAPE Medium to large turtle with a round, flattened carapace covered in sandpapery skin. It also has a long neck and a tubelike snout.

COLOR Gray-green to brown carapace with dark spots. It has green skin with yellow stripes and dark spots.

BEHAVIOR Aquatic. It hauls out on muddy or sandy banks to bask in the sun.

HABITAT Mud- or sand-bottomed rivers, creeks, and lakes.

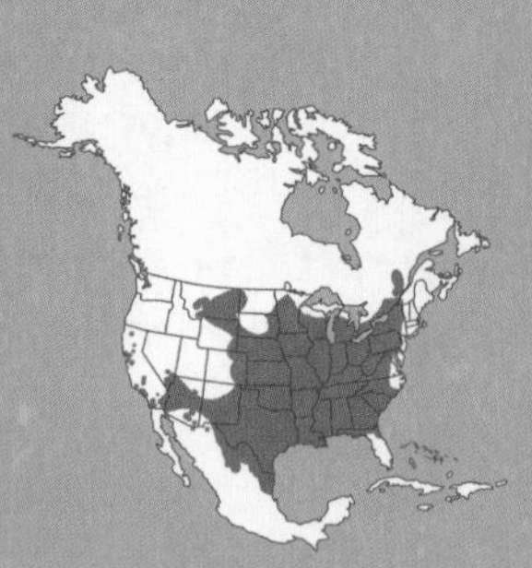

POINT OF FACT Uses its snout like a snorkel to breathe while waiting to ambush crayfish, frogs, or fish.

I SAW IT!

WHEN I SAW IT

DATE

WHERE I SAW IT

SPECIFIC LOCATION, INCLUDE STATE

WHAT IT WAS DOING

NOTES

Diamondback Terrapin

(Malaclemys terrapin)

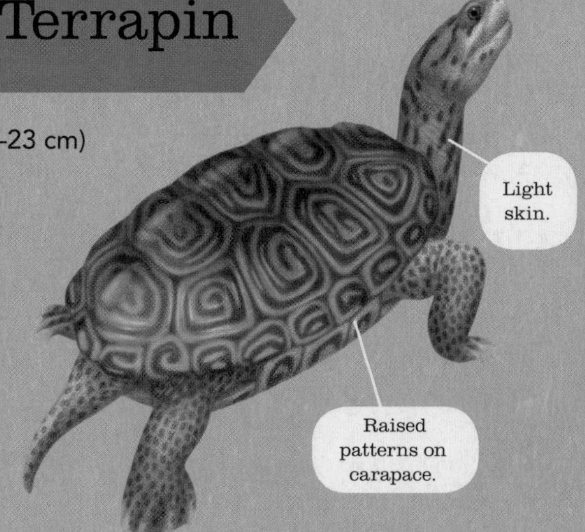

SIZE Shell length 4–9 inches (10–23 cm)

SHAPE Small but thick turtle with ring-like grooves and ridge patterns on its carapace and a large head.

COLOR Brown to black carapace, light gray skin with dark spots and streaks, and a yellow and green plastron.

BEHAVIOR Aquatic. It crunches mussels, crabs, marine snails, and clams with its powerful jaws.

HABITAT Brackish and saltwater coastal marshes, tidal flats, estuaries.

POINT OF FACT Unfortunately, it's threatened by coastal development, boat propellers, pollution, and a history of overhunting for meat. It's even listed as endangered in some states.

I SAW IT!

WHEN I SAW IT
DATE

WHERE I SAW IT
SPECIFIC LOCATION, INCLUDE STATE

WHAT IT WAS DOING

NOTES

Desert Tortoise

(Gopherus agassizii/morafkai)

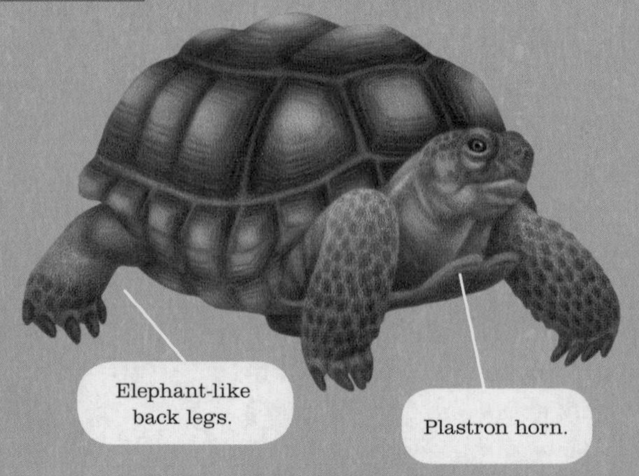

SIZE Shell length 8–15 inches (20½–38 cm)

SHAPE Bulky turtle with a high-domed carapace covered in a raised line pattern. It has thick, stumpy legs covered in large scales, and a long neck.

COLOR Dusty brown to gray carapace with a brown to yellow plastron.

BEHAVIOR Terrestrial. It lives in deep burrows and forages for grasses, cacti, and other plants to eat.

HABITAT Desert, canyons, dunes, washes.

POINT OF FACT The male plastron "horn" sticks out under its chin, which is used to ram into rival males. A female's horn is smaller.

I SAW IT!

WHEN I SAW IT

DATE

WHERE I SAW IT

SPECIFIC LOCATION, INCLUDE STATE

WHAT IT WAS DOING

NOTES

COMMON REPTILES

American Alligator

(Alligator mississippiensis)

Large size.

Rounded snout.

SIZE 6–16 feet (2–5 m) snout to tail tip

SHAPE Huge, low-to-the ground, armor-skinned animal with a long, thick tail and a large mouth full of teeth.

COLOR Dark, gray-brown to black. Young have yellow crossbands

BEHAVIOR Semiaquatic. It basks in the sun near water, digs deep holes in wetlands, and floats with only its nose and eyes above the water's surface.

HABITAT Swamps, lakes, marshes, canals, ponds, rivers.

POINT OF FACT Males bellow and roar. Females bellow and also make pig-like grunts to call their young.

I SAW IT!

WHEN I SAW IT

DATE

WHERE I SAW IT

SPECIFIC LOCATION, INCLUDE STATE

WHAT IT WAS DOING

NOTES

PART V

FISH

What would YOU do?

You're at the beach, enjoying a fun day by the ocean. What better place to find fish! You pull on some flippers, strap on a mask and snorkel, and wade out past the breaking salty surf. While spotting some silvery needlefish, you notice something big and dark under the water off in the distance. What could that be?

You come up to the surface and take off your mask to get a better look. About twenty feet away a fin is cutting through the water—a big, pointy dorsal fin. Uh-oh. Is it a shark? Or is that a dolphin's dorsal fin? Whoever it is seems to be circling around—back toward the beach. Should you swim away or hold still? Could you tell the difference between a dolphin or shark's dorsal fin? *What would you do?*

CHAPTER 1

How to Find Fish

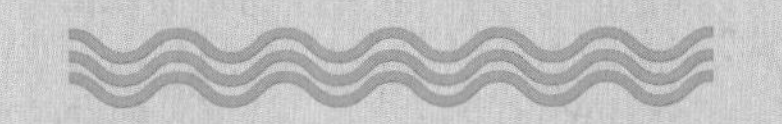

Looking for a lifetime hobby? Fish spotting is a good choice. You could see a new kind of fish every day for the next fifty years and there would still be some out there you hadn't yet seen! There are more species of fish than birds, mammals, reptiles, and amphibians combined. The breadth of fish diversity makes sense if you think about it. The stuff that fish live in—water—makes up most of Earth's surface, after all. Ocean alone covers 70 percent of our watery planet.

Fish (or something very like them) were the first animals with backbones, or vertebrates. They've been swimming around for more than 500 million years. Today you can spot fish everywhere from icy Arctic waters to steamy jungle streams. You can see everything from a dwarf pygmy goby, which is as tiny as your pinkie nail, to a fifteen-ton whale shark as long as a school bus. Now that's diversity!

What Makes a Fish a Fish

Fish live their whole lives in watery oceans, lakes, ponds, rivers, and streams. They even breathe underwater! Fish are the most aquatic of all vertebrate animals, but they need oxygen just like storks, shrews, and snakes do. Instead of breathing in oxygen from air, fish get oxygen from water. The secret is gills. Gills are organs that exchange oxygen and carbon dioxide (like lungs do in land animals) from water.

GILLS

Breathing isn't the only activity water-living fish do differently from land dwellers. Fish navigate in three dimensions.

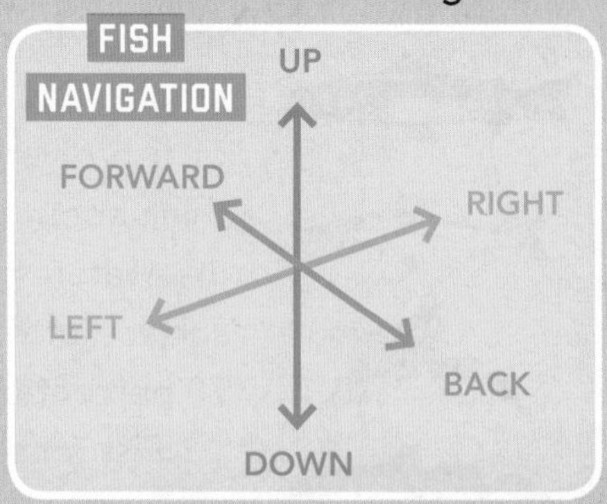

Like you, they go right and left as well as forward and back. But unlike you, they also navigate up and down. (Jump shots don't count! You're coming down, no matter what.) Birds also control their up-and-down position during flight. But water is denser than air. Falling in air and sinking in water aren't the same. The bodies of fish include fins for upward motion, swim bladders, and other adaptations to control buoyancy.

SHARK SKELETON

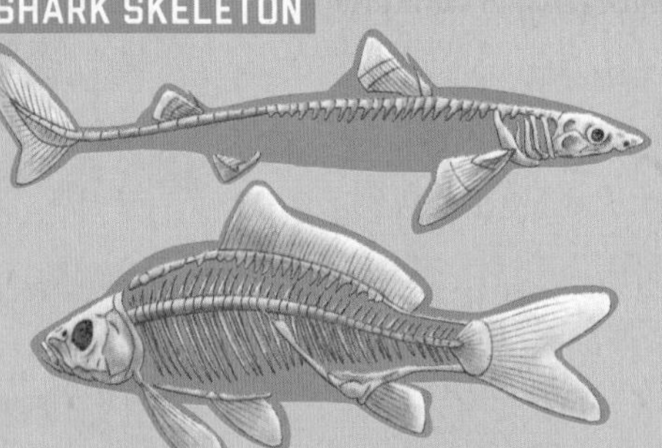

BONY FISH SKELETON

Most fish have bony skeletons, but sharks and rays have flexible skeletons made of cartilage. (The same stuff that makes up your nose.) And like reptiles and amphibians, fish have cold-blooded bodies that change temperature with their surroundings. This means that the water you spot a fish in can tell you a lot about that fish.

Fish Habitats

Brook Trout in cold fresh water.

Channel Catfish in warm fresh water.

Cod in cold salt water.

Tarpon in warm salt water.

Eels are hatched in salt water, then live in fresh water.

Salmon are hatched in fresh water, then live in salt water.

If you've ever snorkeled or gone scuba diving, you have seen firsthand how aquatic habitats change with depth. Like the difference between treetops and forest floor, conditions on the seafloor aren't the same as those near the sunny surface. And if you've ever been fishing, you know that particular fish are adapted to specific habitats. What's catchable in the middle of a lake is different from the fish living in a fast-moving stream. Like all animals, fish have evolved to thrive in certain environments. Some need gravel for egg laying, or current to breathe. And then there's the water itself.

Not all water is the same. Need proof? Stroll the water aisle at the supermarket—but as far as fish go, the two big kinds of water are salt water and fresh water. Most of the world's fishes are saltwater, or marine, fish. There's a lot of ocean out there and it encompasses many kinds of fish habitat—warm, sandy coasts; cold, rocky shores; deep, dark water; sunny tropical reefs; and icy, polar seas. Think

of the variety of temperatures, bottom types, coastal features, depths, marine plants and seaweeds, as well as currents and tides from all over the world in their infinite combinations. That's how many types of marine fish habitat there are, and why there are some 20,000 species of ocean fish. And why finding and watching them is so fascinating.

Freshwater habitats are less varied, but every continent except Antarctica is home to fish that live in lakes, rivers, streams, creeks, marshes, ponds, swamps, and even underground caves. Most fish live in either salt water or fresh water, so the difference is a big clue when identifying the fish you've spotted. Some fish go back and forth between the sea and streams, like salmon, returning to fresh water to spawn. There are also fish that tolerate (or even prefer) semi-salty, brackish water where freshwater rivers empty into the ocean and create coastal swamps and estuaries. They go where the food is—fresh or salty!

TRY IT → Scout a Fish Habitat

Part of spotting and identifying fish is knowing what sort of fish live in a particular place. Sharks don't patrol ponds, and carp don't swim the seas. Get familiar with fish habitats—and the fish that live in them—by surveying one.

STEP 1 **Go to a place where fish live. A beach, creek, pond, lake, etc.**

STEP 2 **As you look around, notice if the water is:**

- **Clear or muddy.**
- **Fast-moving or poky.**
- **Shallow or deep.**
- **Cold or hot.**
- **Fresh or salty.**

STEP 3 **What sorts of fish might live here?**

I DID IT! DATE:

TRACK IT ↘ Fish Habitat Survey

Had a good look around? Time to record the details of your chosen fish-filled place.

WHAT YOU'LL NEED

- a pencil or pen, a jar, metal tape measure or a measuring stick, a waterproof thermometer (like the ones used for an aquarium or pond)

DATE

TIME

LOCATION

WEATHER

HABITAT TYPE:

☐ lake ☐ pond ☐ stream ☐ river ☐ marsh

☐ swamp ☐ beach ☐ tide pool ☐ other

Is there shade (trees, etc.) on the water?

WATER:

☐ fresh ☐ salt ☐ brackish ☐ not sure

How deep (estimate if too deep to measure)?

Color is mostly ☐ clear ☐ green ☐ brown

Anything else?

Deepest visible depth is ....
(use measuring tape or stick)

Check all flow types you can see:

☐ **very fast current (leaf disappears)**

☐ **fast (leaf quickly goes downstream)**

☐ **moderate (leaf slowly goes downstream)**

☐ **slow (seems still)**

Temperature near surface is ....

Temperature near bottom is ....

BOTTOM:

☐ **mud** ☐ **sand** ☐ **gravel** ☐ **rocks**

Is there cover for fish to hide?

☐ **boulders** ☐ **logs** ☐ **water plants**

Did you see any fish?

Other observations?

Could you identify any of the fish that you saw while surveying the fish habitat? Look for them in the Fish Identification section (pages 406–437). Got a positive ID? Check off its I SAW IT! box and fill in the blanks for each species. Congrats on your fabulous fish spotting!

I DID IT! DATE:

TAKE IT TO THE NEXT LEVEL ↗

Nighttime Habitat Check

Can you safely return to your habitat survey site after dark? A whole host of different creatures come out at night—including fish! Use a waterproof flashlight to see who's around, both near the water and in it. A regular flashlight works, too, if you seal it up in a clear jar or zipper-closing baggie. Want to look down deeper? Weight down the baggie or jar with a rock, tie a cord around it, and lower the flashlight into the depths!

I DID IT! DATE:

CHAPTER 2

How to ID Finned Friends

Fish are wildlife, too! And just as with frogs or hawks, identifying fish is about putting these five clues to use.

Here's more detail on what to look for:

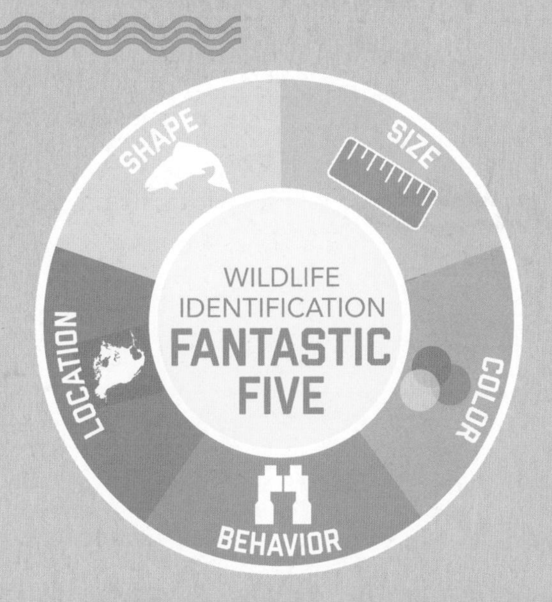

Fish Shapes

Fish vary vastly in shape as well as size. Think of the differences between an eel and angelfish, a shark and stingray, a sea horse and catfish. When describing all those shapes, fish scientists (aka ichthyologists) use some shorthand terms. Let's dive in!

DISC-SHAPED
SUNFISH

Overall body shape is a good place to start. An angelfish is a disc-shaped fish because it's roundish, whereas oval fish are more, well, oval. A fish described as tapering has the shape of a club with one end bigger than the other. Note that these descriptions are of a fish when looking at it from the side. When looking at fish head-on, that angelfish nearly disappears because it's so flattened from side to side—a characteristic

TAPERING
MAHI-MAHI

COMPRESSED
ANGELFISH

scientists call compressed (think pancake). The opposite of compressed is flattened top to bottom, or depressed, like a blobfish. An elongated fish is long for its width, like a sausage, whereas a deep-bodied fish is wide for its width, like a loaf of bread.

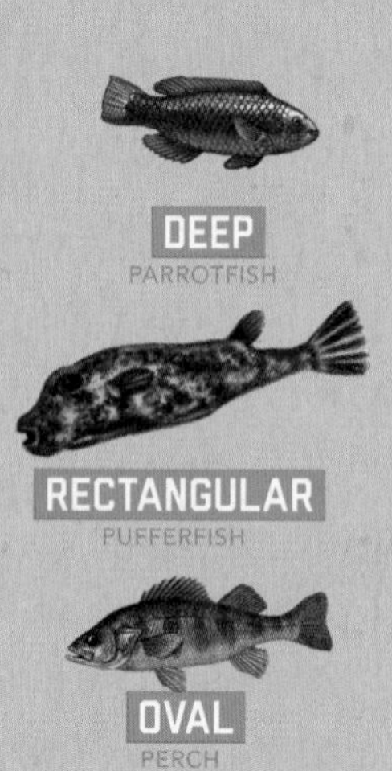

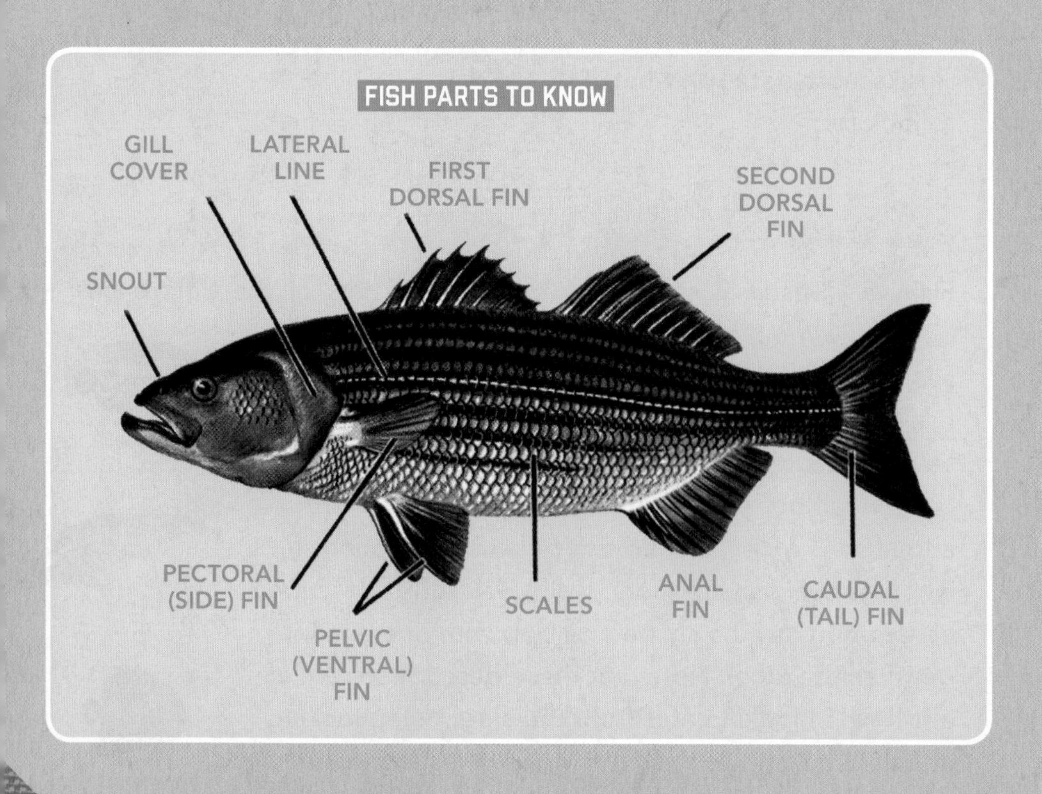

Tails, Fins, and Mouths

Beyond overall body shape and size are the particulars of tails, fins, and mouths. Mouths differ by size but also their position on the fish's head. A mouth on the very end of the snout is *terminal*. On the top of the head a mouth is *superior* and on the bottom it's *inferior*, or ventral. Tail and dorsal fins both come in a variety of shapes and configurations. Their differences make them especially useful for identification.

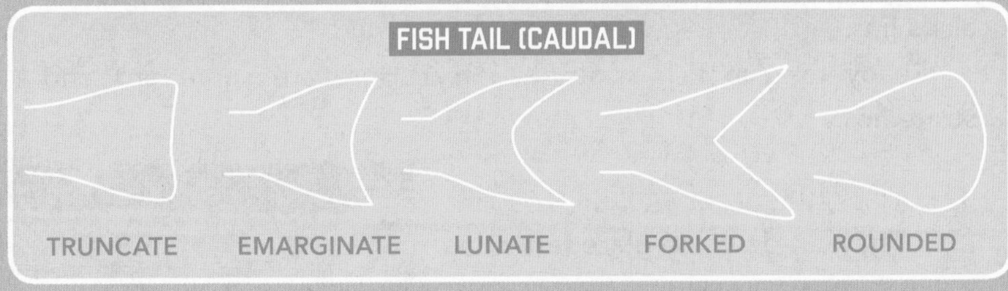

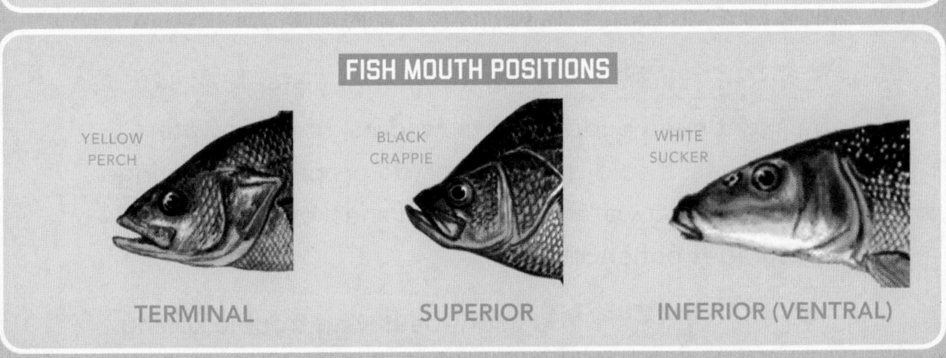

Fish Spotting Tips and Gear

Spotting fish is both easier and harder than seeing other kinds of wildlife. Since fish live in water, you're limited in the places to look. Unlike searching for birds or foxes in a local park, the only place to find fish is in the park's pond or creek. On the other hand, it can be hard to spot fish in deep or murky water. And while field guides feature identification images that show a fish on its side, if you're keeping dry it's likely you're looking down at a fish, so you can't see all its fins.

Ready to give fish spotting a try? Here are some tips and strategies:

WEAR POLARIZED SUNGLASSES They cut down on glare reflecting off the water's surface.

SEEK THE SHADE Water shaded by trees or clouds is easier to see into.

FIND AN ANGLE Looking directly down into water is often the least revealing. Crouch down or move around when looking into shallow water along seashores or clear streams. Find the angle with the clearest view.

DRESS FOR SUCCESS If the weather or water is chilly, you might need puddle boots, waders, or other water-proof boots. If cold isn't a factor, do prepare for getting wet by wearing water shoes or old sneakers. Discomfort discourages patience!

STILL THE WATERS You might be having trouble seeing fish, but they see you! Once you're in a spot, stop and quietly wait. Remaining still will encourage whoever swam away to return.

GET SOME GEAR Ichthyologists (fish scientists to you and me) and people who catch fish for sport use a bathyscope (also called an aquascope) when fish hunting. A bathyscope is an underwater viewer and many are simply a cone or tube with a clear plastic bottom that you hold in the water and look through. Having water up against the clear panel makes a viewing window, kind of like looking into an aquarium. (Glass-bottomed boats do the same thing.) You can make your own, too. (See page 404.)

Aquascopes vary, but all give a better view of what's underwater.

GET WET Holding a diving mask in the water works like a bathyscope. And swimming around while wearing a mask or goggles is even better. Know how to snorkel? Go for it!

CATCH AND RELEASE Seeing fish in their environment is great, but what if you want to get a closer look? If you've got a net and quick reflexes, consider catching a fish. Bring along a net and a clear jar. Fill the jar with water from the creek, shore, or pond to keep the fish in while you observe it. Be gentle and return the fish where you found it.

HUNT AT NIGHT Nocturnal fish can be extra difficult to spot. One way to catch a glimpse is to hold a waterproof flashlight (or a flashlight in a zipper-closing baggie) underwater and see who's around. (See page 392.)

Stripes, spots, and other patterns on fish are often useful field marks.

CRITTER CONFUSION: Shark or Dolphin?

SHARK

A shark's dorsal fin is more of a triangle, with a straight edge on its back side (the side hopefully moving away from you).

DOLPHIN

A dolphin's dorsal fin is curved with a narrower tip. Think it's a shark? Don't panic and start wildly swimming away or splashing and thrashing your arms and legs. That makes you chase worthy. Keep your eyes on the shark, move away slowly, and make your way back to the boat or shore. Get out of the shark's path to the open ocean.

Fish Families

Each fish species page in the Fish Identification section (pages 406–437) includes the family it belongs to next to an icon that represents the overall shape of its members. These aren't the official scientific family names, but are instead the groups of fishes in the family. For example, the puffer fish family is called Tetraodontidae, but we're just calling it Puffers to ease brain fatigue.

By knowing who's in which fish family, you can get a better sense of the differences in fish shapes. Take a look at the members of the Minnows family (Cyprinidae) in the Fish Identification section: Fathead Minnow (page 423), Common Carp (page 411), and Golden

Shiner (page 422). While they are different in size and other features, all three do have a similar body shape. Now find the five Sunfish family members in the Fish Identification section. How are they similar and/or different?

Talk Like an Ichthyologist

Here are some terms to know when reading field guides and talking fish fins and tails.

ADIPOSE FIN a small fleshy fin without rays

ANADROMOUS a fish (like salmon) that hatches in fresh water, migrates to the ocean, then returns to fresh water to spawn

ANGLER someone who catches fish with a fishing rod

BARBEL a thin, fleshy structure near mouth for finding food

BUOYANCY the amount something floats or sinks

BRACKISH water that is somewhat or slightly salty

CAUDAL FIN tail fin

CRUSTACEAN crayfish, pill bug, shrimp, crab, or other invertebrate animal with hard outer shells and jointed legs

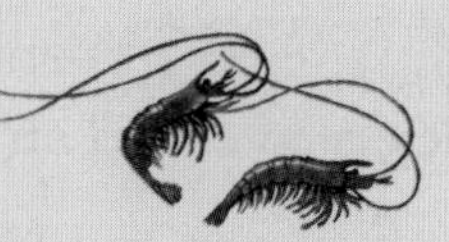

SHRIMP

DORSAL top or upper side

ESTUARY brackish body of water that connects to the sea and has a river flowing into it

FIELD MARK distinctive feature useful in identification

FRESH WATER water with little or no salt

FRY recently hatched baby fish

GILL organ for breathing in water

GILL COVER bone covering gills

ICHTHYOLOGIST person who studies fishes

INVERTEBRATES animals without backbones, including worms, bugs, crayfish, etc.

RAYS

LATERAL LINE a canal of pores along a fish's body with sense organs

RAY soft, flexible structures that support fins

SCHOOL a group of fish of the same species

SPAWNING fish breeding activity

SUBSTRATE bottom of a lake, stream, sea, or pond

SWIM BLADDER an air-filled organ that regulates buoyancy, also called an air bladder

TEMPERATE moderate, between warm and cool

VENTRAL bottom side

VERTEBRATE an animal with a backbone

TRY IT → Find Fish

Ready to put your fish identification skills into practice? Go where fish are and starting spotting them! (There are some tips on page 396.)

WHAT YOU'LL NEED

- appropriate clothing, a net and jar (optional)

STEP 1 Find some wild fish to watch in the water, or catch one to observe in a jar.

STEP 2 As you look at the fish, check out its:

- Fins: Can you pick out the pectoral, pelvic, and anal fins? What's the dorsal fin like?
- Tail: What's its shape?
- Mouth: Size and shape are things to look for.
- Color: Stripes, spots, patterns can be notable field marks.

STEP 3 As you observe the fish, think about what its *most* distinctive characteristics—its field marks—are. Is it color or pattern? An enormous mouth or oddly shaped fin?

STEP 4 Can you identify the fish? Use the Fish Identification section (pages 406–437) and/or a field guide to help. If it's in the Fish Identification section, check off the I SAW IT! box on its page and fill in the blanks. Congratulations!

TRACK IT ↘ Tracking Fish

Now that you've spent some quality time with a fish, record what you observed.

WHAT YOU'LL NEED

- a pencil or pen

BLUEGILL

DATE

TIME

LOCATION

WEATHER

Describe habitat

Estimate length of body from snout to end of tail:

Do you think it's an adult?

☐ yes ☐ no ☐ not sure

Overall shape of the fish's body (Check all that apply):

☐ compressed ☐ depressed ☐ elongated

☐ disc-shaped ☐ tapering ☐ fusiform

☐ rectangular ☐ oval

Shape of tail fin:

☐ forked ☐ truncate ☐ rounded ☐ other

Shape of dorsal fin:

☐ single continuous ☐ two separate fins

☐ other

Is the dorsal fin:

☐ spiny ☐ soft ☐ both spiny and soft

Shape of pectoral (side) fins:

Shape of pelvic (ventral) fins:

Is the pelvic fin nearer the

☐ anal fin ☐ pectoral fin?

Estimate length of mouth (include units)

Color

Stripes, spots, lines, or other patterns?

What would you say are its field marks?

What's it doing?

How does it move?

Is it solitary or in a group?

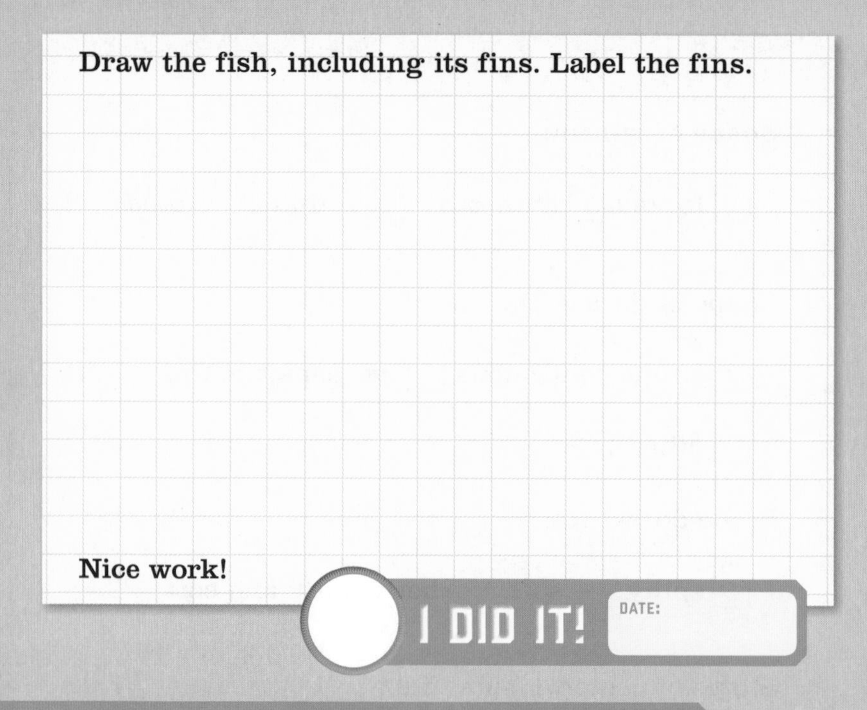

TAKE IT TO THE NEXT LEVEL

Make a Bathyscope

A bathyscope is a pretty simple device when you think about it. Pushing water up against clear plastic or glass creates a viewing window into the water. Anything that accomplishes this will help you see underwater. Got a mixing bowl, flower vase, or salad bar container with a clear, flat, see-through bottom? Stick the flat end into the water and see for yourself!

Want to make a dedicated device? One with shaded sides to cut down on glare even more? Try this one.

WHAT YOU'LL NEED

- **a length of wide drainpipe, coffee or other wide can, or juice jug; plastic wrap or clear plastic sheeting; rubber bands; waterproof tape; scissors**

STEP 1 Make a waterproof cylinder out of pipe, or a can or plastic container or tub with ends removed. It doesn't have to be round! Dig around in the recycling bin to find something. Sharp or jagged edges? Cover them with tape.

STEP 2 Place the cylinder on the plastic wrap or sheeting. Cut around it with 2–3 inches (5-7½ cm) of extra plastic.

STEP 3 Stretch the cut piece of plastic over one end of the cylinder and secure with rubber bands.

STEP 4 Wrap tape around the rubber bands and cover the end of the plastic wrap to keep out water.

STEP 5 Place the plastic-covered end in the water and look through the open end. Who do you see?

FISH IDENTIFICATION

Welcome to your guide to fish identification. Here are some tips to get started and where in the book (page numbers) you'll find more info:

FRESHWATER FISH

Freshwater fish are listed first, then saltwater fish.

The fish's family name and an icon of the fish family's overall shape are included.

INCHES 1

Note that lengths are averages. Fish size varies a lot!

Blue edged gill cover.
Flat, round body.

The labeled main illustration points out field marks, including mouth positions and fin shapes. Take note!

Habitat tells you the water temperature, as well as the bodies of water, the fish lives in.

Check the range map to see where the fish lives.

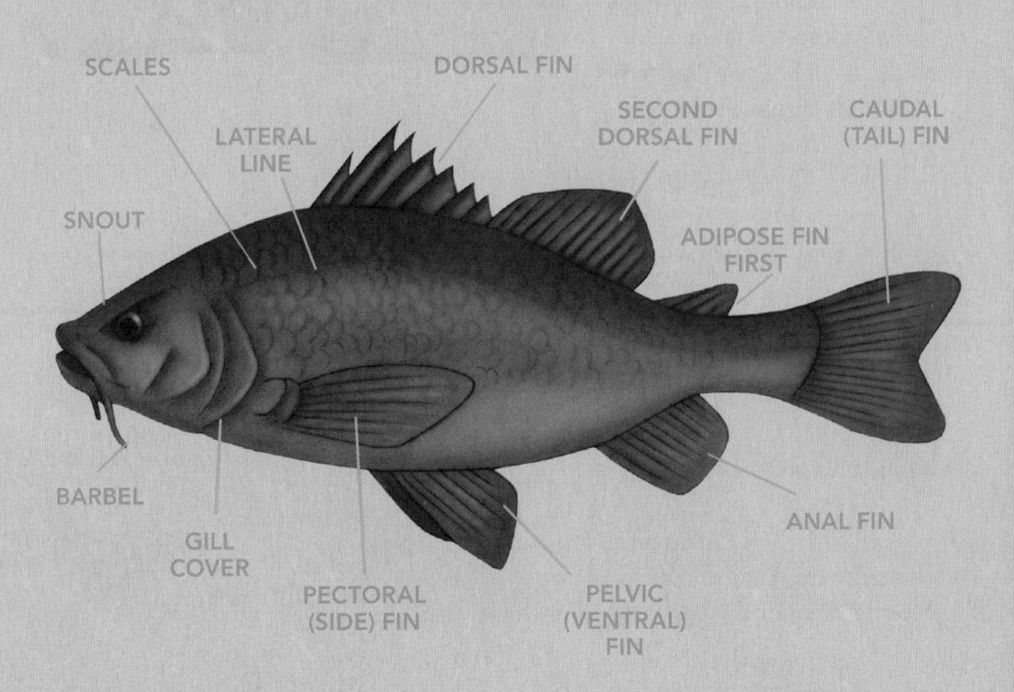
SCALES
DORSAL FIN
SECOND DORSAL FIN
CAUDAL (TAIL) FIN
LATERAL LINE
SNOUT
ADIPOSE FIN FIRST
BARBEL
GILL COVER
PECTORAL (SIDE) FIN
PELVIC (VENTRAL) FIN
ANAL FIN

FRESHWATER FISH

Channel Catfish

(Ictalurus punctatus)

FAMILY Catfishes

SHAPE Slender, smooth-skinned fish with large eyes near the top of its wide head and a forked tail fin. It also has a small, fleshy adipose fin behind its single dorsal fin.

AVERAGE LENGTH 12–20 inches (30½–51 cm)

COLOR Gray back, a white belly, and light gray to silvery sides with small dark spots.

BEHAVIOR It looks for food at night and spends its days in deep water.

HABITAT Warm water. Large streams with deep pools, plus lakes and ponds.

POINT OF FACT It eats everything including algae, plants, snails, fish, snakes, and frogs. Taste buds all over its slimy body help it to sense food nearby.

I SAW IT!

WHEN I SAW IT
DATE

WHERE I SAW IT
SPECIFIC LOCATION, INCLUDE STATE

WHAT IT WAS DOING

NOTES

FRESHWATER FISH

Bluegill

(Lepomis macrochirus)

FAMILY Sunfishes

SHAPE Round, flat, saucer-shaped body with eyes on its sides, a small mouth, and a forked tail. Its single dorsal fin is joined by spiny and soft segments.

AVERAGE LENGTH Up to 16 inches (40½ cm)

COLOR Dark green back, silver to yellow sides, faint, vertical bars, and a large dark spot on the rear of its dorsal fin. Breeding males have bright red-orange bellies.

BEHAVIOR Lives in schools of 10–20 fish. It spends mornings near warm, sunny water surfaces and seeks deep shade or weeds on hot summer days.

HABITAT Warm water. Quiet streams, ponds, swamps, weedy lake edges, medium rivers.

POINT OF FACT It's one of the most popular fish caught by anglers for fun or sport, and it's stocked in ponds for bass food.

I SAW IT!

WHEN I SAW IT

DATE

WHERE I SAW IT

SPECIFIC LOCATION, INCLUDE STATE

WHAT IT WAS DOING

NOTES

FRESHWATER FISH

Green Sunfish

(Lepomis cyanellus)

FAMILY Sunfishes

SHAPE Flat but thick, with a stretched, oval-shaped body and a short, rounded pectoral fin. Its single dorsal fin is joined by spiny and soft segments.

AVEARAGE LENGTH 4–6 inches (10–15 cm), but up to 12 inches (30½ cm)

COLOR Dark green back with sides that fade from blue to yellow. It also has yellow-orange edged fins and a white belly.

BEHAVIOR Males build nests in shallow, weedy water for females to lay eggs in.

HABITAT Warm water. Quiet pools in slow streams, lakes, and ponds.

POINT OF FACT It survives in cloudy water with low oxygen and other potentially harmful conditions, making it one of the most common sunfishes.

I SAW IT!

WHEN I SAW IT

DATE

WHERE I SAW IT

SPECIFIC LOCATION, INCLUDE STATE

WHAT IT WAS DOING

NOTES

....

....

FRESHWATER FISH

Common Carp

(Cyprinus carpio)

FAMILY Minnows

SHAPE Thick body with a flat belly and a single, long-arched dorsal fin. It also has a down-turned mouth and a forked tail fin.

AVEARAGE LENGTH 16–18 inches (40½–45½ cm) but up to 47 inches (119½ cm)

COLOR Brassy yellow to light brown back and sides with a white belly.

BEHAVIOR Uses its snout to dig around in the mud for food and gulps for air at the water's surface.

HABITAT Warm water. Shallow quiet streams, muddy river pools, lakes, ponds.

POINT OF FACT People have been farming pond carp for a thousand years, and humans introduced the fish to waterways around the world—including North America—over a century ago.

I SAW IT!

WHEN I SAW IT

DATE

WHERE I SAW IT

SPECIFIC LOCATION, INCLUDE STATE

WHAT IT WAS DOING

NOTES

FRESHWATER FISH

Brook Trout

(Salvelinus fontinalis)

FAMILY Salmons

SHAPE Bullet-shaped body, a small, fanlike dorsal fin, and a down-turned mouth. It has a small, fleshy adipose fin behind its single dorsal fin.

AVERAGE LENGTH 10 inches (25½ cm)

COLOR Dark green to gray back with wavy cream-colored lines, light, splotchy sides, and a red belly and fins.

BEHAVIOR Females dig out nests in gravel, then cover their eggs once they are fertilized.

HABITAT Cold water. Clear streams and rivers, small gravel-bottomed lakes.

POINT OF FACT It's the only trout that's native to eastern North America. It's also a popular sport fish that's often called a "brookie" or "speckled trout."

I SAW IT!

WHEN I SAW IT
DATE

WHERE I SAW IT
SPECIFIC LOCATION, INCLUDE STATE

WHAT IT WAS DOING

NOTES

FRESHWATER FISH

Rainbow Trout

(Oncorhynchus mykiss)

FAMILY Salmons

SHAPE Bullet-shaped body with a down-turned mouth and large eyes. It has a small, fleshy adipose fin behind its fan-like dorsal fin.

AVEARAGE LENGTH 12–22 inches (30½–56 cm), larger in lakes

COLOR Blue-green head and back with silver to white lower sides and belly. It has a pink band or stripe down its sides and small black spots all over.

BEHAVIOR It uses water plants, sunken logs, and boulders as underwater cover while hunting smaller fish and crustaceans.

HABITAT Cold water. Clear streams and creeks, rivers, and lakes.

POINT OF FACT Those that migrate from streams into the open ocean are called "steelhead" because of their silver heads.

I SAW IT!

WHEN I SAW IT
DATE

WHERE I SAW IT
SPECIFIC LOCATION, INCLUDE STATE

WHAT IT WAS DOING

NOTES

....

....

FRESHWATER FISH

Black Crappie

(Pomoxis nigromaculatus)

Dorsal spines.

Dip in arch over eye.

Large mouth.

FAMILY Sunfishes

SHAPE Flat, stretched, oval-shaped body with an arched back, equal-size dorsal and anal fins, a round, forked tail fin, and a large mouth with an underbite. Its single dorsal fin is joined by spiny and soft segments.

AVERAGE LENGTH 7–12 inches (18–30½ cm), but up to 19 inches (48½ cm)

COLOR Black to dark green back, blotchy silver sides, and dark fins with lighter splotches.

BEHAVIOR Nests in colonies and gathers in schools to feed on aquatic insects and smaller fish.

HABITAT Warm water. Lakes, ponds, slow stream backwater pools with plants.

POINT OF FACT Males guard nests for five days until their eggs hatch and juveniles start feeding on their own.

I SAW IT!

WHEN I SAW IT

DATE

WHERE I SAW IT

SPECIFIC LOCATION, INCLUDE STATE

WHAT IT WAS DOING

NOTES

FRESHWATER FISH

Largemouth Bass

(Micropterus salmoides)

FAMILY Sunfishes

SHAPE Long, oval, thick body with a forked tail fin, a rounded pectoral fin, and a large, down-turned mouth that extends beyond its eye. Its single dorsal fin is joined by spiny and soft segments.

AVERAGE LENGTH 12–20 inches (30½–51 cm), but up to 29 inches (73½ cm)

COLOR Dark green back and green sides, often with a dark band along a lateral line, a white belly, and brown eyes.

BEHAVIOR Males guard over nests until juveniles leave home.

HABITAT Warm water. Lakes, ponds, swamps, rivers, and creeks with mud or sandy bottoms.

POINT OF FACT It's often the top predator of its aquatic habitat, because it eats whatever moves—including streamside mice and birds!

I SAW IT!

WHEN I SAW IT
DATE

WHERE I SAW IT
SPECIFIC LOCATION, INCLUDE STATE

WHAT IT WAS DOING

NOTES

....

....

FRESHWATER FISH

Smallmouth Bass

(Micropterus dolomieu)

Red eye.

Spiny segments in front.

Brown bands.

FAMILY Sunfishes

SHAPE Long thick, oval-shaped body with a forked tail fin, a rounded pectoral fin, and a down-turned mouth that extends beyond its eye. Its single dorsal fin is joined by spiny and soft segments.

AVERAGE LENGTH 12–20 inches (30½–51 cm)

COLOR Mottled, dark green sides with bronze speckles, dark brown vertical bands, a white belly, and red eyes.

BEHAVIOR Males build and guard nests that are often near logs or boulders for shelter.

HABITAT Warm water. Clear river pools or rocky lakes with gravel bottoms.

POINT OF FACT It likes cooler, clearer water than the largemouth bass, but it also tolerates warmer, cloudier water than trout. So it has replaced them in many streams.

I SAW IT!

WHEN I SAW IT

DATE

WHERE I SAW IT

SPECIFIC LOCATION, INCLUDE STATE

WHAT IT WAS DOING

NOTES

FRESHWATER FISH

Walleye

(Sander vitreus)

Brassy flecks.

Large teeth.

White tail tip.

FAMILY Perches

SHAPE Long, slender body that ends in a forked tail fin. It has a pointy snout, a huge mouth with sharp teeth, silver eyes, and two separate dorsal fins.

AVERAGE LENGTH 14–17 inches (35½–43 cm), but up to 36 inches (91½ cm)

COLOR Dark, brassy yellow to brown back and sides, a white belly, and a dark spot at the base of its spiny dorsal fin.

BEHAVIOR Spawns in groups that chase each other, swim in circles, and display with fins to others.

HABITAT Temperate water. Lakes, clear pools in backwaters, and rivers.

POINT OF FACT A reflective layer of pigment in its eye allows it to see in low light and hunt at night, dawn, dusk, and on dark, cloudy days.

I SAW IT!

WHEN I SAW IT

DATE

WHERE I SAW IT

SPECIFIC LOCATION, INCLUDE STATE

WHAT IT WAS DOING

NOTES

....

....

FRESHWATER FISH

Yellow Perch

(Perca flavescens)

FAMILY Perches

SHAPE Long oval body with two separate dorsal fins, forked tail fin, and a large mouth.

AVERAGE LENGTH 8–11 inches (20½–28 cm), but up to 16 inches (41 cm)

COLOR Brown-yellow with 6–9 dark vertical bars down its sides, orange lower fins, and a white belly.

BEHAVIOR Feeds on insects, snails, leeches, crayfish, and small fish.

HABITAT Temperate water. Clear lakes, ponds, creek and river pools near vegetation.

POINT OF FACT Females camouflage their eggs by laying them in long ribbons within underwater plants and branches in the springtime.

I SAW IT!

WHEN I SAW IT

DATE

WHERE I SAW IT

SPECIFIC LOCATION, INCLUDE STATE

WHAT IT WAS DOING

NOTES

....

....

FRESHWATER FISH

Longnose Gar

(Lepisosteus osseus)

Plate-like scales.

FAMILY Gars

SHAPE Long, skinny, tube-shape body with a snout twice as long as its head, needlelike teeth, a single dorsal fin near its rump, and a rounded tail fin.

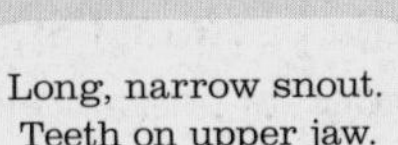

AVERAGE LENGTH 12–36 inches (30½–91½ cm), but up to 72 inches (183 cm)

COLOR Dark green to black with dark spots on its sides.

BEHAVIOR Feeds mostly at night on small fish like minnows.

HABITAT Temperate water. Slow pools and backwaters of rivers, lakes, some brackish water.

POINT OF FACT Its modified swim bladder aids delivery of air to its gills, allowing it to breathe in cloudy, low-oxygen water and even out of water for a short while.

I SAW IT!

WHEN I SAW IT

DATE

WHERE I SAW IT

SPECIFIC LOCATION, INCLUDE STATE

WHAT IT WAS DOING

NOTES

....

....

FRESHWATER FISH

Johnny Darter

(Etheostoma nigrum)

FAMILY Perches

SHAPE Small, slender, and long body with a rounded snout, large pectoral fins, two separate dorsal fins, and a truncate tail fin.

AVERAGE LENGTH 2–3 inches (5–7½ cm)

COLOR Blotchy, mottled beige with dark X and W patterns on its sides.

BEHAVIOR It darts and sinks in the water, swimming with a quick sprint and dropping to the bottom to navigate the rushing current.

HABITAT Temperate water. Flowing creeks, small to medium rivers, sandy lake shores.

POINT OF FACT Its darting motion—it is a darter!—makes it hard to see in moving waters, but look for them perched in place on their pectoral fins.

I SAW IT!

WHEN I SAW IT

DATE

WHERE I SAW IT

SPECIFIC LOCATION, INCLUDE STATE

WHAT IT WAS DOING

NOTES

....

....

FRESHWATER FISH

Mosquitofish

(Gambusia affinis)

FAMILY Livebearers

SHAPE Club-shaped body with a wide chest, a flattened head, and an upturned mouth. It has a single rounded dorsal fin and a rounded tail fin. Males have a long anal fin with a pointy tip. Females are larger than males.

AVERAGE LENGTH 1½–3 inches (4–7½ cm)

COLOR Dull gray to beige with dark teardrop markings below its eyes.

BEHAVIOR Males transfer sperm to females using their anal fin's tubes. Females give birth to live young.

HABITAT Warm and temperate water. Slow to standing water in ponds, ditches, and weedy stream pools.

POINT OF FACT It's a commonly stocked gobbler of aquatic mosquito larvae that's able to live in warm, low-oxygenated water.

I SAW IT!

WHEN I SAW IT

DATE

WHERE I SAW IT

SPECIFIC LOCATION, INCLUDE STATE

WHAT IT WAS DOING

NOTES

FRESHWATER FISH

Golden Shiner

(Notemigonus crysoleucas)

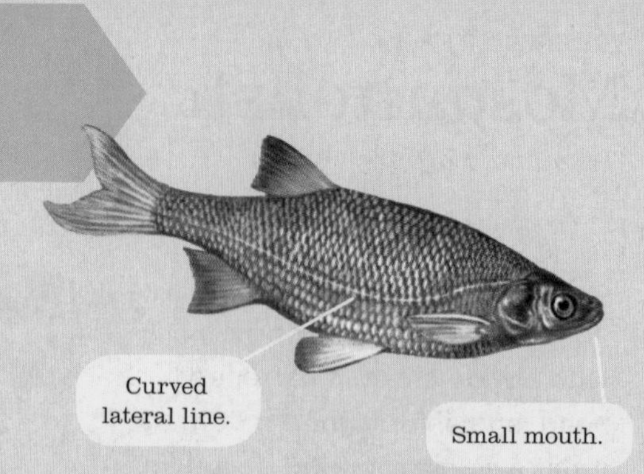

FAMILY Minnows

SHAPE Almond-shaped body with a small upturned mouth, a forked tail fin, and a fanlike dorsal fin.

AVERAGE LENGTH 5–6 inches (12½–15 cm), but up to 12 inches (30½ cm)

COLOR Green-gold with a dusky line. Southern populations have red fins.

BEHAVIOR Hides from its many predators among underwater weeds.

HABITAT Temperate and warm water. Weedy lakes, ponds, swamps, and slow pools in creeks and rivers.

POINT OF FACT It's bred and sold as baitfish to anglers, which has caused it to be introduced far beyond its native range.

I SAW IT!

WHEN I SAW IT
DATE

WHERE I SAW IT
SPECIFIC LOCATION, INCLUDE STATE

WHAT IT WAS DOING

NOTES

FRESHWATER FISH

Fathead Minnow

(Pimephales promelas)

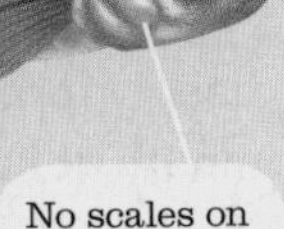

FAMILY Minnows

SHAPE Long, flat body with a short, blunt head, a small mouth, eyes on the sides of its head, a single, rounded dorsal fin, and a forked tail. Spawning males have bumps on their snouts.

AVERAGE LENGTH 3–4 inches (7½–10 cm)

COLOR Dark green to blue-gray back with yellow sides, dark side stripes, and a lighter belly.

BEHAVIOR Males make nests under branches or rocks; then females swim upside down to attach their sticky eggs as they're laid.

HABITAT Temperate water. Muddy creeks and small rivers, ponds.

POINT OF FACT It's tolerant of hot, low-oxygen, cloudy water, so it survives in dry creek pools and low summer ponds.

I SAW IT!

WHEN I SAW IT

DATE

WHERE I SAW IT

SPECIFIC LOCATION, INCLUDE STATE

WHAT IT WAS DOING

NOTES

....

....

FRESHWATER FISH

Gizzard Shad

(Dorosoma cepedianum)

FAMILY Herrings

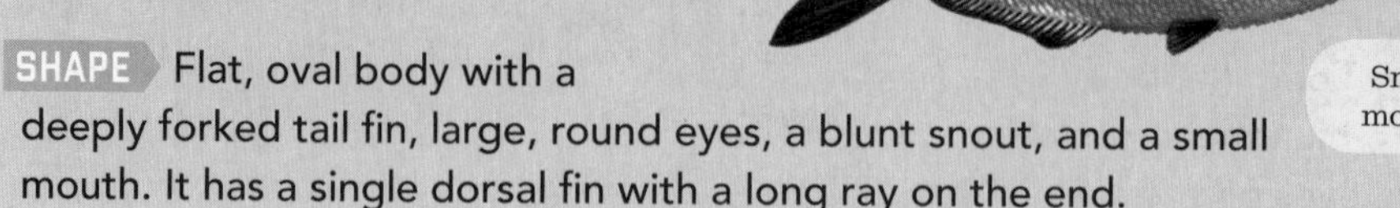

SHAPE Flat, oval body with a deeply forked tail fin, large, round eyes, a blunt snout, and a small mouth. It has a single dorsal fin with a long ray on the end.

AVERAGE LENGTH 10–15 inches (25½–38 cm), but up to 20 inches (51 cm)

COLOR Silver, pale-blue back with white sides and a white belly. Younger fish have dark spots behind their gills.

BEHAVIOR Swims around during the day to filter tiny plants and animals out of the water.

HABITAT Temperate fresh water and will enter brackish water. Open water in rivers, lakes, and reservoirs.

POINT OF FACT Its name comes from its long, gizzard-like, tangled, and sand-filled intestines.

I SAW IT!

WHEN I SAW IT
DATE

WHERE I SAW IT
SPECIFIC LOCATION, INCLUDE STATE

WHAT IT WAS DOING

NOTES

....

....

FRESHWATER FISH

Skipjack Herring

(Alosa chrysochloris)

FAMILY Herrings

SHAPE Long, almond-shaped body with a single dorsal fin, a deeply forked tail fin, and a large mouth.

AVERAGE LENGTH 12–16 inches (30½–40½ cm), but up to 20 inches (51 cm)

COLOR Blue-green back and silver sides with no spots.

BEHAVIOR It leaps out of the water while chasing small fish and insects, quickly skipping across the surface.

HABITAT Temperate water. Open water of larger rivers and reservoirs, often over sand or gravel bottoms.

POINT OF FACT Endangered elephant-ear and ebony shell mussels attach to and live on skipjack herring as larva. These freshwater mussels are endangered in waters where skipjack herring have disappeared.

I SAW IT!

WHEN I SAW IT
DATE

WHERE I SAW IT
SPECIFIC LOCATION, INCLUDE STATE

WHAT IT WAS DOING

NOTES

....

....

FRESHWATER FISH

Northern Pike

(Esox lucius)

FAMILY Pikes 

SHAPE Long, tubular body with a flat head, a big mouth, and a forked tail fin. It has a single dorsal fin over its anal fin.

AVERAGE LENGTH 18–24 inches (45½–61 cm), but up to 56 inches (142 cm)

COLOR Dark green back, green sides, and rows of irregular oval, yellow spots.

BEHAVIOR It uses its back-slanting teeth to ambush and hunt large fish.

HABITAT Cool to temperate water. Clear, weedy lakes, slow rivers and creeks.

POINT OF FACT Crossbreeding the Northern pike and muskellunge fish creates "tiger muskie," which are popular with anglers that fish for fun.

I SAW IT!

WHEN I SAW IT

DATE

WHERE I SAW IT

SPECIFIC LOCATION, INCLUDE STATE

WHAT IT WAS DOING

NOTES

....

....

FRESHWATER FISH

White Sucker

(Catostomus commersonii)

FAMILY Suckers

SHAPE Long, tubular body with a rounded snout, a large mouth with no teeth, and thick lips. It has one angled dorsal fin and a forked tail fin.

AVERAGE LENGTH 12–18 inches (30½–45½ cm), but up to 25 inches (63½ cm)

COLOR Green to brown back with gray to silver sides and a cream belly.

BEHAVIOR It forages along the bottom of the water by sucking up bits of plant, insects, and crustaceans.

HABITAT Temperate water. Rocky pools of streams, lakes, rivers.

POINT OF FACT It's also called a black mullet, bay fish, coarse-scaled sucker, and eastern sucker.

I SAW IT!

WHEN I SAW IT
DATE

WHERE I SAW IT
SPECIFIC LOCATION, INCLUDE STATE

WHAT IT WAS DOING

NOTES

FRESHWATER/SALTWATER FISH

American Eel

(Anguilla rostrata)

Continuous combined dorsal, tail, and anal fin.

Snake-like shape.

Large mouth.

FAMILY Eels

SHAPE Smooth, snakelike body with a small, pointed head, a large mouth, and a long dorsal fin that extends beyond half of its body.

AVERAGE LENGTH 24–36 inches (61–91½ cm), but up to 60 inches (152½ cm)

COLOR Dark brown back with yellow sides and a white belly.

BEHAVIOR Hunts for insects, small fish, and crayfish at night and hides from predators during the day.

HABITAT Temperate water. Flowing inland streams, brackish tidewater areas.

POINT OF FACT Mature adults return to Sargasso Sea in the North Atlantic Ocean where they spawn and then die. Females lay up to 20 million eggs over their lifetimes.

I SAW IT!

WHEN I SAW IT

DATE

WHERE I SAW IT

SPECIFIC LOCATION, INCLUDE STATE

WHAT IT WAS DOING

NOTES

FRESHWATER/SALTWATER FISH

Chinook Salmon

(Oncorhynchus tshawytscha)

FAMILY Salmons

SHAPE Thick, oval body with a large, truncated tail fin, a large mouth, and small eyes. It has one angled dorsal fin and a small, fleshy adipose fin.

AVERAGE LENGTH 24–30 inches (61–76 cm), but up to 58 inches (147 cm)

COLOR Metallic green to blue-green upper sides and back, silver lower sides, and black dots on its back and tail. Breeding males are dark green-brown to purple.

BEHAVIOR Migrates from the sea (or lake where it was introduced) up streams to spawn.

HABITAT Cold water. Coastal streams, ocean, cold lakes where introduced.

POINT OF FACT It's big and meaty, so it often weighs in at 20–30 pounds (9–14 kg) with some reaching over 120 pounds (54 kg)!

I SAW IT!

WHEN I SAW IT
DATE

WHERE I SAW IT
SPECIFIC LOCATION, INCLUDE STATE

WHAT IT WAS DOING

NOTES

SALTWATER FISH

Spiny Dogfish

(Squalus acanthias and *Squalus suckleyi)*

FAMILY Dogfish Sharks

SHAPE Long, slender, flat-bellied body with a pointed snout, huge eyes, gill slits, and rough skin. Its two dorsal fins are spaced far apart, and the upper lobe of its forked tail fin is larger than the lower lobe.

AVERAGE LENGTH 31–40 inches (78½–101½ cm), but up to 5 feet (1½ m)

COLOR Gray or brown top with white spots along its back and a lighter belly.

BEHAVIOR Hunts in packs (like wild dogs!) for squid, fish, crab, jellyfish, and other invertebrates.

HABITAT Temperate coastal ocean waters along shores over sandy or soft seafloor.

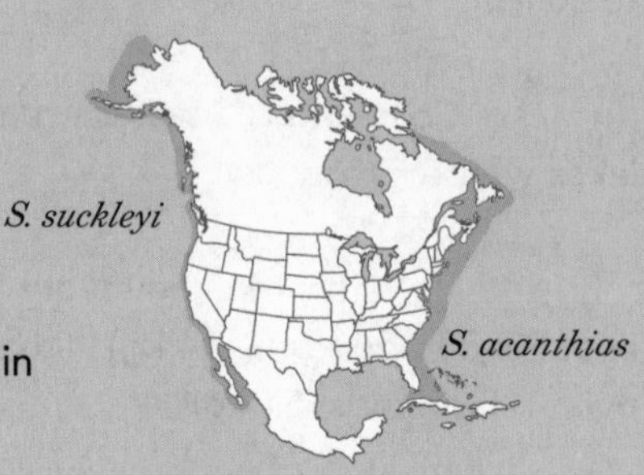

POINT OF FACT It's often the fish in "fish & chips" in Europe, but overfishing has made this abundant, harmless shark less common in recent years.

I SAW IT!

WHEN I SAW IT
DATE

WHERE I SAW IT
SPECIFIC LOCATION, INCLUDE STATE

WHAT IT WAS DOING

NOTES

SALTWATER FISH

Summer Flounder

(Paralichthys dentatus)

Eyes on left.

Separate tail fin.

Small pelvic fin.

FAMILY Sand Flounders

SHAPE Flat, disc-shaped body with both of its eyes on the left side of its head. Its only dorsal fin runs along its entire back, and it has an anal fin along its belly.

AVERAGE LENGTH 15–20 inches (38–51 cm) but up to 37 inches (94 cm)

COLOR Brown with eye-shaped spots on the top half of its body.

BEHAVIOR It hides by burrowing all but its eyes into seafloor mud or sand, then surprises passing prey with a big chomp.

HABITAT Soft bottoms, often in bays and near shore.

POINT OF FACT It eats small fish and invertebrates, and spends the winter in deep water and the summer in the shallows to spawn.

I SAW IT!

WHEN I SAW IT

DATE

WHERE I SAW IT

SPECIFIC LOCATION, INCLUDE STATE

WHAT IT WAS DOING

NOTES

....

....

SALTWATER FISH

Smooth Butterfly Ray

(Gymnura micrura)

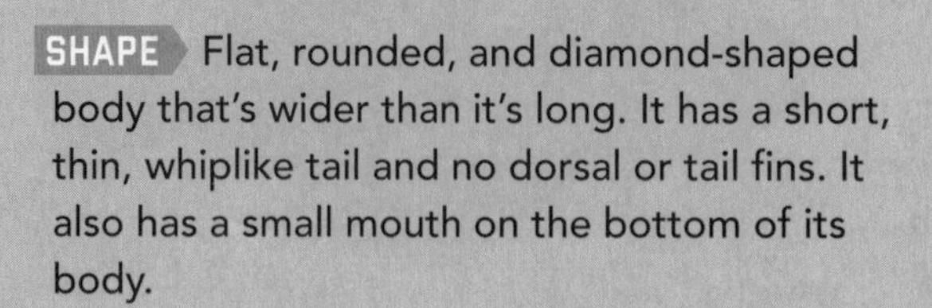

FAMILY Butterfly Rays

SHAPE Flat, rounded, and diamond-shaped body that's wider than it's long. It has a short, thin, whiplike tail and no dorsal or tail fins. It also has a small mouth on the bottom of its body.

AVERAGE WIDTH 16–20 inches (41–50 cm), but up to 36 inches (91 cm)

COLOR Gray, brown, light green, or purple with dot and line markings and white underneath.

BEHAVIOR It changes color with its surroundings to hide from hammerhead sharks and other predators.

HABITAT Shallow temperate waters over sandy bottoms.

POINT OF FACT Rays, skates, and sharks have skeletons made of cartilage, not bone. They make up the cartilaginous fishes group, separate from regular bony fishes.

I SAW IT!

WHEN I SAW IT
DATE

WHERE I SAW IT
SPECIFIC LOCATION, INCLUDE STATE

WHAT IT WAS DOING

NOTES

....

....

SALTWATER FISH

Northern Puffer

(Sphoeroides maculatus)

FAMILY Puffers

SHAPE Club-shaped, narrow body with a flat belly, a large head, huge eyes, and a tiny mouth. It has short, wispy dorsal and anal fins, and no pelvic fin.

AVERAGE LENGTH 8–10 inches (20½–25½ cm)

COLOR Brown to gray back with dark blotches, black spots, and dark bars on its sides, and a yellow to white belly.

BEHAVIOR When threatened, it quickly inflates itself with water or air into a prickly ball to warn off enemies.

HABITAT Bays and estuaries along coast in summer, offshore in winter.

POINT OF FACT Two tough teeth in each of its jaws create a beak-like mouth that can crack into shellfish for a meal.

I SAW IT!

WHEN I SAW IT

DATE

WHERE I SAW IT

SPECIFIC LOCATION, INCLUDE STATE

WHAT IT WAS DOING

NOTES

SALTWATER FISH

Tidepool Sculpin

(Oligocottus maculosus)

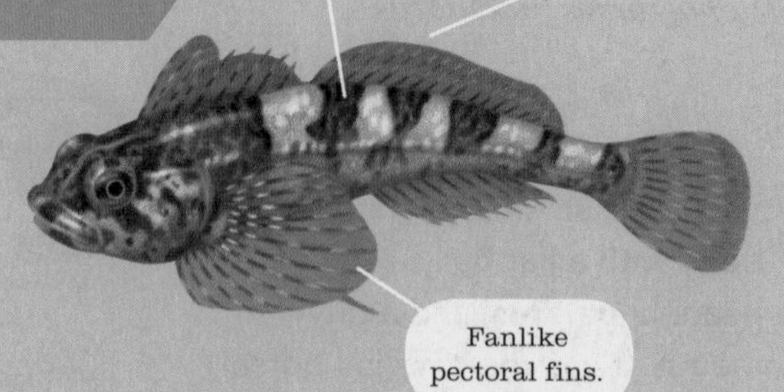

FAMILY Sculpins

SHAPE Small, tapered body with a large head, bulging eyes, and wide, fanlike pectoral fins. It has two separate dorsal fins, one of which is spiny.

AVERAGE LENGTH 3½ inches (9 cm)

COLOR Green to rusty brown variations with a white to blue belly.

BEHAVIOR Hunts worms and small crustaceans in sheltered spots along coast.

HABITAT Cold to temperate water. Tide pools along rocky shores, shallow coastal areas, under rocks in intertidal zones.

POINT OF FACT It depends on camouflage and speed, more than its spines, to avoid predators—just like other kinds of sculpins.

I SAW IT!

WHEN I SAW IT
DATE

WHERE I SAW IT
SPECIFIC LOCATION, INCLUDE STATE

WHAT IT WAS DOING

NOTES

....

....

SALTWATER FISH

Shiner Perch

(Cymatogaster aggregata)

Yellow bars on sides. Long dorsal fin.

Small mouth.

FAMILY Surfperches

SHAPE Flat, oval-shaped body with a single, long dorsal fin, a forked tail fin, and a long anal fin.

AVERAGE LENGTH 4–5 inches (10–12½ cm), but up to 7 inches (18 cm)

COLOR Light green back, silver sides, and three vertical, faded-yellow bars on its sides with patches of dark, spotty rows in between.

BEHAVIOR Feeds on zooplankton in water and forages on its seabed.

HABITAT Cold-temperate water. Bays, rocky shores, kelp beds, coastal waters over soft seabeds.

POINT OF FACT It's also called a "shiner surfperch." This common fish is often caught by anglers fishing off piers.

I SAW IT!

WHEN I SAW IT

DATE

WHERE I SAW IT

SPECIFIC LOCATION, INCLUDE STATE

WHAT IT WAS DOING

NOTES

SALTWATER FISH

Sheepshead

(Archosargus probatocephalus)

FAMILY Porgies

SHAPE Large head and a high-backed body with a mouth full of blunt teeth. It has a long pectoral fin, a forked tail fin, and one long, spiny dorsal fin.

AVERAGE LENGTH 10–20 inches (25½–51 cm), but up to 36 inches (91½ cm)

COLOR Silver with dark vertical bands along body.

BEHAVIOR Feeds in groups and chomps on oysters, clams, and crabs with its tough mouth and strong teeth.

HABITAT Temperate to warm water. Shallow water over oyster beds, near piers and bridges.

POINT OF FACT In the Great Lakes region, the name "sheepshead" refers to a freshwater drum fish *(Aplodinotus grunniens)* and isn't related to this saltwater species!

I SAW IT!

WHEN I SAW IT
DATE

WHERE I SAW IT
SPECIFIC LOCATION, INCLUDE STATE

WHAT IT WAS DOING

NOTES

SALTWATER FISH

Atlantic Needlefish

(Strongylura marina)

Long, narrow mouth.

Stripe on side.

Light-colored fins.

FAMILY Needlefishes 

SHAPE Long, extremely narrow body with long, toothy jaws. It has a single dorsal fin far down its back and a truncated tail fin. Its anal fin is larger than its dorsal fin.

AVERAGE LENGTH 9–11 inches (23–28 cm), but up to 25 inches (63½ cm)

COLOR Green to turquoise above and silver below. It has a thin, dark stripe down each of its sides.

BEHAVIOR Like tweezers, it grabs small fish from the side with its long jaws, then gulps them down.

HABITAT Temperate and warm water. Coastal waters, mangrove lagoons, and freshwater streams near shore.

POINT OF FACT Females hide their eggs by attaching them to seaweed and other floating objects with long, sticky threads.

I SAW IT!

WHEN I SAW IT
DATE

WHERE I SAW IT
SPECIFIC LOCATION, INCLUDE STATE

WHAT IT WAS DOING

NOTES

....

....

101 OUTDOOR SCHOOL

ANIMAL WATCHING ACHIEVEMENTS

The outdoors are calling—and this is the list of everything they have to offer. Track everything you master, experience, and collect on your adventures by checking off your achievements below. See if you can complete all 101!

- 1 Spotted a raptor hunting
- 2 Found a nest with bird eggs in it
- 3 Identified a bird by its song
- 4 Used binoculars
- 5 Spot an endangered animal
- 6 Found a wild bird feather
- 7 See turtles hatch
- 8 Found an animal tooth
- 9 Spotted bird scat
- 10 Spotted reptile scat
- 11 Found a snake skin
- 12 Watch an animal play dead
- 13 Found an animal bone
- 14 Spotted a bird with an eyebrow stripe
- 15 Spotted an animal hole
- 16 Spotted a molehill
- 17 Watch an animal stalk its' prey
- 18 Learned my local venomous snakes
- 19 Spotted a bird with wing bars
- 20 Caught a frog (carefully)
- 21 Hear a frog or toad call at night
- 22 Watch a badger dig a home
- 23 Identified an LBJ bird
- 24 Spotted a bird with spectacles
- 25 Watch a beaver build a home
- 26 Spotted a vagrant bird
- 27 Identified a bird by its silhouette
- 28 Drew a bird using two circles
- 28 Spotted a tadpole
- 30 Identified a juvenile bird
- 31 Imitate a bird song and attract birds
- 32 Saw a bird catch a bug in flight
- 33 Spotted a hummingbird
- 34 Saw a hawk gliding in the sky
- 35 Found an abandoned bird nest
- 36 Spotted a saltwater fish
- 37 Saw a bird feeding chicks
- 38 Found a bird's eggshell
- 39 Spotted a bird dust bathing
- 40 See a turtle snap
- 41 Heard a woodpecker pecking
- 42 Spotted a squirrel carrying food
- 43 Spotted crows mobbing a hawk

- 44 Identified a bird in flight
- 45 Found a stick chewed by a beaver
- 46 Identified a game trail
- 47 Spotted a bat at night
- 48 Watch an animal eat its prey
- 49 Spotted a duck dabbling
- 50 Spotted a heron fishing
- 51 Identified a freshwater fish
- 52 Spotted birds courting
- 53 Heard an owl at night
- 54 Found bird tracks
- 55 Identified a black bear (safely)
- 56 Found a bird's cavity nest
- 57 Spotted a salamander
- 58 Spotted a newt
- 59 Identified a frog by its call
- 60 Identified toad eggs
- 61 Spotted a tail drag
- 62 Found wild animal fur
- 63 Identified mammal herbivore scat
- 64 Heard a squirrel chattering
- 65 Heard coyotes yipping at night
- 66 Spotted claw marks on a tree
- 67 Identified a woodchuck burrow
- 68 Spotted a leafy squirrel nest in a tree
- 69 Smelled skunk scent
- 70 Identified a tail feather
- 71 Found an antler
- 72 Identified a mammal by its tracks
- 73 Spotted deer tracks
- 74 Found chewed-on plants, nuts, or cones
- 75 Heard an animal whistle
- 76 Identified an animal from eyeshine
- 77 Drew a mammal using three circles
- 78 Spotted tree bark ribboned by antlers
- 79 Drew an animal track
- 80 Identified mammal carnivore scat
- 81 Heard a deer huff
- 82 Spotted a reptile basking
- 83 Found a turtle shell
- 84 Found a reptile eggshell
- 85 Identified a female bird
- 86 Identified a uric pellet
- 87 Found a vernal pond or pool
- 88 Spotted a snake with crossbands
- 89 Identified a snake track
- 90 Spotted frog tracks
- 91 Found salamander eggs
- 92 Heard toads calling
- 93 Identified a lizard
- 94 Spotted a freshwater fish
- 95 Spotted a fish with a forked tail fin
- 96 Spotted a bird building a nest
- 97 Identified a dolphin by its fin
- 98 Identified pectoral, pelvic, and anal fins on a fish
- 99 Found a fish skeleton
- 100 Identified a saltwater fish
- 101 Spotted a fish with a notched dorsal fin

INDEX

alligator, 310, 322–23, 381
amphibians and reptiles, 340–57, 360–81
  identification tips, 315–29, 338–39, 358–59
  scouting/tracking, 304–14, 330–37
angelfish, 393
armadillo, 202, 224, 275

badger, 207, 222–23, 226, 259, 438
bass, 395, 415–16
bat, 33, 215, 227, 233–35, 239
  big brown, 299
  eastern red, 298
  free-tailed, 223, 300
  hoary, 301
bear, 2, 191, 196, 208, 227, 248
beaver, 196, 218, 225–26, 232, 237, 274
binoculars, 16–18, 22–23
birds, 88–187
  behaviors, 56–77, 86
  identification tips, 36–41, 47–55, 86–87
  scouting/tracking, 26–35, 42–46, 78–85
blackbird, 63, 101, 106, 158
bluebird, 47, 54, 62, 168, 177
bluegill, 402, 409
bluehead, 388
bobcat, 196, 207, 224, 226, 252
bobolink, 161
bobwhite, 169
bunting, indigo, 27, 47, 63, 72

cardinal, 36, 47, 49, 63, 75, 96
carp, 411
catbird, gray, 73, 92
catfish, 387, 408
cautionary tips, 10–11, 196, 326
cedar waxwing, 120
chickadee, 62, 78, 104, 185
chipmunk, 8, 196, 202, 205, 209, 234, 280–82
cod, 387
coot, 153
cormorant, 147
cottontail, 215, 268
coyote, 196, 206–07, 226, 251
crane, 60, 152
crappie, black, 395, 414
crossbill, 36, 126
crow, 42, 44–45, 57, 59, 107
cyprinodon, 389

deer, 9, 172, 196, 206–07, 216–17, 226, 236, 238, 242–43
dickcissel, 162
dipper, 176
dogfish shark, 430
dolphin, 207, 222, 234, 256, 384, 398
dove, 39, 95, 111
duck, 39, 58, 70, 94, 140

eagle, 49, 60, 65, 79, 146, 174
eel, 387, 394, 428
egret, 150
elk, 215–16, 225–26, 244

finch, 39, 59, 72, 121–22, 184
  goldfinch, 18, 28, 48, 50, 58–9, 89
fish, 408–37
  finding, 384–92
  identification tips, 393–407
flicker, 97
flounder, 431
flycatcher, 58
fox, 196, 207, 225–26, 236, 249–50
frog, 305, 307–8, 315–17, 331–33, 338–39
  bullfrog, 351
  green, 340
  northern leopard, 320, 350
  spring peeper, 315, 347
  tree, 318, 333, 345–46
  western chorus, 337, 348
  wood, 349

gambusia, 389
gar, longnose, 419
gear and tools, 16–19, 22–23, 193, 396–97
gecko, 311
gnatcatcher, 73
gopher, 285–86
grackle, 106
grosbeak, 129
grouse, 61, 180
grunt fish, 398
gull, 38, 40, 66, 100

harrier, 165
hawk, 27–28, 39, 58–9, 65, 148, 151, 165
  Harris's, 109
  red-shouldered, 142
  red-tailed, 102
heron, 47, 58–9, 65, 80, 144, 150, 156
herp. See amphibians and reptiles
herring, 424–25
hummingbird, 27, 38, 56, 65, 103, 112

jackrabbit, 267
jay, 3, 47, 58, 62, 88, 175
  blue jay, 47, 58, 62, 88
  Steller's jay, 3, 175
junco, 123

kestrel, 171
killdeer, 66, 167
killifish, 389
kingfisher, 43, 65, 80, 143
kinglet, 186

lark, horned, 164
lizard, 305–6, 311, 322, 330, 333, 358–59
  collared, 328, 369
  eastern fence, 312, 370
  greater short-horned, 373
  green anole, 372
loon, 80

magpie, 170
mahi-mahi, 393
mallard, 94
mammals, 242–301
  identification tips, 202–14, 222–41
  scouting/tracking, 191–201, 215–21
marmot, 270
meadowlark, 163
mink, 207, 262
minnow, 411, 422–23
mockingbird, 20, 99
mole, 215, 218, 223, 236, 288–89
moose, 190, 207, 226, 246
mosquitofish, 421
mountain lion, 7, 253
mouse, 227, 290–92, 294
mudpuppy, 357
muskrat, 222, 271

needlefish, 437
newt, 318, 352–53
nutcracker, 181
nuthatch, 28, 56, 183

opossum, 202, 206, 227, 233, 261
oriole, 66, 131
osprey, 151
otter, 218, 222, 226, 272
ovenbird, 63, 134
owl, 39–40, 57, 62, 65, 127–28

parrotfish, 388, 394
pearl razor fish, 388
pelican, 43, 145
perch, 394–95, 417–18, 420, 435
pigeon, 60, 98
pigfish, 398
pike, 426
porcupine, 223–24, 226, 265
prairie dog, 223, 276
pronghorn, 247
ptarmigan, 173
puffer, 394, 398, 433

quail, 116, 169

rabbit, 205–7, 215, 224, 226, 267–68
raccoon, 196, 202, 206, 227, 260
rat, 295–97
raven, 59, 178
reef fish, 388

reptiles. See amphibians and reptiles
ringtail, 223, 263
roadrunner, 43, 113
robin, 26, 42, 44–45, 52, 62, 66–67, 90

salamander, 305–8, 311–13, 315–318, 331–33, 338–39, 357, 371
  red-backed, 311, 356
  spotted, 355
  tiger, 307, 318, 354
salmon, 387–88, 394, 399, 429
sapsucker, 138
sculpin, 434
sea lion, 203, 207, 209, 255
seal, 192, 207, 215, 224, 254
shad, gizzard, 424
shark, 384, 386, 398, 430
sheep, bighorn, 225, 245
sheepshead fish, 436
shrew, 203, 224, 227, 287
shrike, 36
shrimp, 399
skink, 323, 371
skunk, 196, 224, 227, 264, 266
snake, 304, 310, 322, 325, 330, 358–59
  coachwhip, 365
  coral, 323, 326
  eastern hog-nosed, 360
  garter, 368
  gopher, 362
  king, 364
  rat, 331, 366
  rattle, 326, 361
  ringneck, 363
  water, 367
sparrow, 42, 44–45
  black-throated, 118
  field, 28, 172
  house, 57, 93
  song, 28, 105
  white-crowned, 187
  white-throated, 119
squirrel, 227
  flying, 206, 283
  ground, 203–4, 284–85
  tree, 200, 203, 207, 236, 277–79
starling, 91
stingray, 394
sucker, 395, 427
sunfish, 393, 410
swallow, 29, 58, 149, 160
swift, 33, 66, 108

tanager, 47, 130, 179
tern, 80, 154
thrasher, 117
thrush, 36, 63, 136
titmouse, 28, 63, 125
toad, 8, 316–17, 331–32, 338–39
  American, 318, 333, 341
  eastern spadefoot, 344
  great plains, 343
  Woodhouse's, 342
tools. See gear and tools
trout, 387, 412–13
turkey, wild, 141
turtle, 312, 322, 324, 330–31, 358–59
  box, 305, 310, 374
  desert tortoise, 380
  diamondback terrapin, 379
  painted, 377
  red-eared slider, 309, 375
  snapping, 376
  spiny softshell, 378

verdin, 115
vole, 293
vulture, turkey, 166

walleye, 395, 417
warbler, 61, 63, 78, 132–33, 157
weasel, 226, 273
whale, 207, 234, 257–58
woodchuck, 224, 226, 236, 269
woodpecker, 38, 59, 61, 97, 114, 137, 139, 182
wood-pewee, 135
wren, 62, 110, 124

yellowlegs, 155
yellowthroat, 36, 63, 159

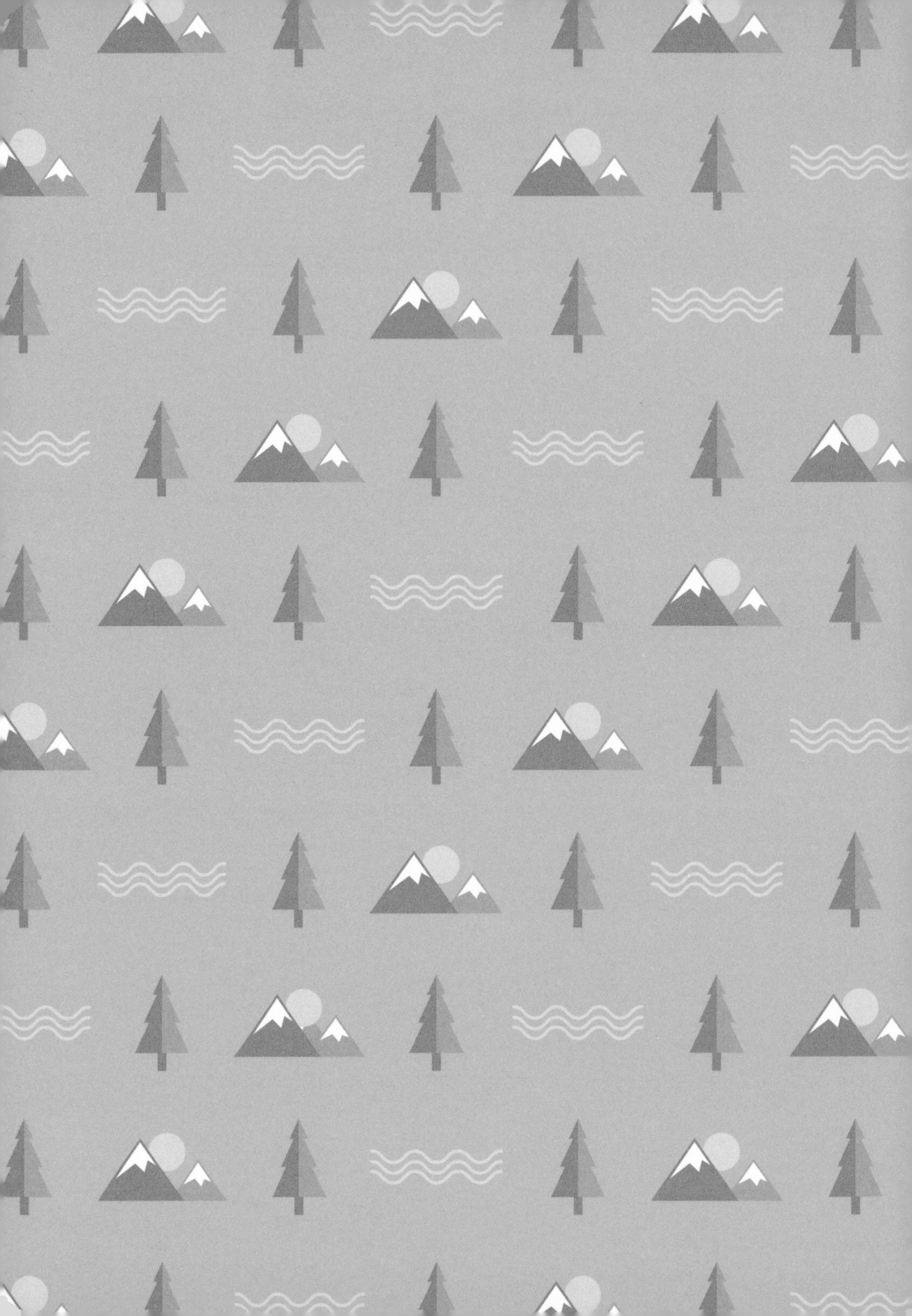